Michael Brooks

13 THINGS THAT DON'T MAKE SENSE

Michael Brooks, Ph.D., is formerly senior features editor, and now a consultant for *New Scientist*, in which the wildly popular article on which this book is based first appeared. His writing has appeared in the *Guardian*, the *Independent*, and the *Observer*. He lives in England.

www.13thingsthatdontmakesense.com

13 THINGS THAT DON'T MAKE SENSE

13 THINGS THAT DON'T MAKE SENSE

THE MOST BAFFLING SCIENTIFIC MYSTERIES OF OUR TIME

Michael Brooks

Vintage Books

A Division of Random House, Inc.

New York

This book is based on an article that originally appeared in the March 19, 2005, issue of *New Scientist*.

The Library of Congress has cataloged the Doubleday edition as follows:
Brooks, Michael, 1970–
13 things that don't make sense : the most baffling scientific mysteries of our time /
Michael Brooks.
p. cm.
Includes bibliographical references and index.
1. Science—Miscellanea. II. Title.
Q173.B893 2008
500—dc22 2008012443

Vintage ISBN: 978-0-307-27881-4

Author photograph © Andrew Perris
Book design by Elizabeth Rendfleisch

www.vintagebooks.com

Printed in the United States of America
10 9 8

*To Mr. Sumner, for lasting inspiration and fascination.
I hope this repays some of my debt.*

Also to Phillippa, Millie, and Zachary for inspiration every day.

The most exciting phrase to hear in science, the one that heralds the most discoveries, is not "Eureka!," but "That's funny . . ."

—ISAAC ASIMOV

CONTENTS

13 THINGS THAT DON'T MAKE SENSE

PROLOGUE

I am standing in the magnificent lobby of the Hotel Metropole in Brussels, watching three Nobel laureates struggle with the elevator.

It's certainly not an easy elevator to deal with; it's an open mesh cage, with a winch system that looks like something Isambard Kingdom Brunel might have built. When I first got into it three days ago, I felt like I was traveling back in time. But at least I got it to work.

Embarrassed for the scientists, I look away for a moment and distract myself with the grandeur of my surroundings. The Metropole was built at the end of the nineteenth century and is almost ridiculously ornate. The walls are paneled with vast slabs of marble, the ceilings decorated in subtle but beautiful gold and sage green geometric patterns. The glittering crystal chandeliers radiate a warmth that makes me want to curl up and go to sleep beneath their light. In fact, there are glowing, comforting lights everywhere. Outside, in the Place de Brouckère, the wind is blowing a bitter cold across the city; faced with the bleak December beyond those revolving doors, I feel like I could stand here forever.

The Nobel laureates are still struggling. No one else seems to have noticed their plight, and I'm wondering whether to walk across the lobby and offer help. When I had my long fight with the door, I discovered there's something about the shutter mechanism that defies logic—when you think it must be locked, it isn't; it needs a final pull. But it occurs to me that people who have attached Nobel Academy pins to their lapels ought to be able to work that out for themselves.

I like to think of scientists as being on top of things, able to explain the world we live in, masters of their universe. But maybe that's just a comforting delusion. When I can tear myself away from the farce playing out in the elevator, I will be getting into a cab and leaving behind perhaps the most fascinating conference I have ever attended. Not because there was new scientific insight—quite the contrary. It was the fact that there was no insight, seemingly no way forward for these scientists, that made the discussions so interesting. In science, being completely and utterly stuck can be a good thing; it often means a revolution is coming.

The discussion at the conference was focused on string theory, the attempt to tie quantum theory together with Einstein's theory of relativity. The two are incompatible; we need to rework them to describe the universe properly, and string theory may be our best bet. Or maybe not. I have spent the last three days listening to some of today's greatest minds discuss how we might combine relativity and quantum theory. And their conclusion was that, more than three decades after the birth of string theory, we still don't really know where to start.

This was a Solvay physics conference, a meeting with the richest of histories. At the first Solvay conference in 1911—the world's first physics conference—the delegates debated what was to be made of the newly discovered phenomenon of radioactivity. Here in this hotel Marie Curie, Hendrik Lorentz, and the young Albert Einstein debated how it was that radioactive materials could apparently defy the laws of conservation of energy and momentum. Radioactivity was an anomaly; it didn't make sense. The problem was eventually solved by the birth of quantum theory. At the 1927 Solvay conference, though, the strange nature of quantum theory caused its own problems, provoking Einstein and Niels Bohr, Lorentz and Erwin Schrödinger, Ernest Rutherford and John von Neumann to sit discussing these new laws of physics with the same degree of confusion as they had shown toward radioactivity.

It was an extraordinary moment in the history of science. Quantum theory encapsulated the novel idea that some things in nature are entirely random, happen entirely without cause. This made no sense to Einstein or Bohr, and the pair spent their time outside the formal discussions sparring over what it all meant. They had entirely different philosophical approaches to dealing with that mystery, however. To Bohr, it meant some things might be beyond the scope of science. To Einstein, it meant something was wrong with the theory; it was here in this hotel that Einstein made his famous remark that "God does not play dice." Bohr's reaction faces up to scientists' biggest frustration: that they don't get to set the rules. "Einstein," he said, "stop telling God what to do."

Neither man lived to see a satisfactory solution to the enigma—it remains unsolved, in fact. But if some delegates at the twenty-third Solvay conference are to be believed, it seems Bohr might have been right about there being limits to science. Half of the string theorists present, some of the greatest minds in the world, are now convinced that we can never fully comprehend the universe. The other seekers after a "theory of everything" think there must be some explanation available to us. But they have no idea where to find it. What has led to this extraordinary situation? Yet another anomaly.

This one was discovered in 1997. Analysis of the light from a distant supernova led astronomers to a startling conclusion: that the universe is expanding, and that this expansion is getting faster and faster all the time. The revelation has stunned cosmologists; no one knows why this should be so. All they can say is that some mysterious "dark energy" is blowing up the universe.

This anomaly, an apparently simple observation, has brought string theory to its knees. It cuts away at everything its proponents thought they had achieved. Put simply, they can't explain it—and many of them feel they should stop trying. There is a straightforward answer staring us in the face, they say: our universe must be one of many universes, each with different characteristics. To try to find reasons why those characteristics are as they are in our universe, they argue, is a waste of time.

But it is not. There is something inspiring about this—and any—anomaly. When Thomas Kuhn wrote *The Structure of Scientific Revolutions* in the early 1960s, he wanted to examine the history of science for clues to the nature of discovery. The clues led him to invent the term—now a cliché—*paradigm shift*. Scientists work with one set of ideas about how the world is. Everything they do, be it experimental or theoretical work, is informed by, and framed within, that set of ideas. There will be some evidence that doesn't fit, however. At first, that evidence will be ignored or sabotaged. Eventually, though, the anomalies will pile up so high they simply cannot be ignored or sabotaged any longer. Then comes crisis.

Crisis, Kuhn said, is soon followed by the paradigm shift in which everyone gains a radically new way of looking at the world. Thus were conceived ideas like relativity, quantum theory, and the theory of plate tectonics.

The dark energy situation is another such crisis. You can see it as depressing, a hint that science has hit a brick wall. But, equally, you can see it as exciting and inspiring. Something has now got to give, and the breakthrough could come from anywhere at any time. What is even more exciting is that it is not the only anomaly of our time—not by a long way.

It is not even the only one in cosmology. Another cosmic problem, dark matter, was first spotted in the 1930s. Following Kuhn's template almost exactly, it was ignored for nearly forty years. Vera Rubin, an astronomer at Washington, D.C.'s Carnegie Institution, was the one to nail it down and make people deal with it. In the early 1970s, she showed that the shape, size, and spin of galaxies means either there is something wrong with gravity or there's much more matter out there in space than we can see. No one wants to mess with Newton's laws governing gravity, but neither do we know what this dark matter might be.

It's sometimes comforting to imagine that science is mastering the universe, but the facts tell a different story. Put together, dark matter and dark energy make up 96 percent of the universe. Just two anomalous scientific results have told us that we can see only a tiny fraction of what we call the cosmos. The good news is that cosmologists are now, perhaps, emerging from Kuhn's crisis stage and are in the process of reinventing our universe—or they will be once they manage to work out where the paradigm shift should lead.

Other, equally stirring anomalies—revolutions-in-waiting, perhaps—await our attention closer to home too. There is the placebo effect: carefully planned, rigorously controlled experiments repeatedly show that the mind can affect the body's biochemistry in ways that banish pain and produce startling medical effects. Except that, like dark matter, no one is quite sure that the placebo effect really exists. Cold fusion experiments, where nuclear reactions inside metal atoms safely release more energy than they consume, have also survived nearly two decades of skepticism, and the U.S. Department of Energy recently declared that the laboratory evidence is strong enough to merit funding of a new round of experimental research. The thing is, cold fusion goes against all the received wisdom in physics; there is no good explanation for why it should work—or even strong evidence that it does. But it is still worth investigating: the hints that we do have suggest that it could expose a new, deeper theory of physics that could have an enormous impact on many aspects of science. Then there is the "intelligent" signal from outer space that has defied explanation for thirty years; the enigma of our sense of free will despite all scientific evidence to the contrary; the spacecraft that are being pushed off course by an unknown force; the trouble we have explaining the origin of both sex and death using our best biological theories . . . the list goes on.

The philosopher Karl Popper once said, rather cruelly perhaps, that "science may be described as the art of systematic oversimplification." Though that is an oversimplification in itself, it is clear that science still has plenty to be humble about. But here is the point that is often missed by scientists eager to look as if nothing is beyond their abilities. Dark energy has been described as the most embarrassing problem in physics. But it is not; it is surely the greatest opportunity in physics—it gives us reason to examine our oversimplifications and correct them, bringing us to a new state of knowledge. The future of science depends on identifying the things that don't make sense; our attempts to explain anomalies are exactly what drives science forward.

In the 1500s, a set of celestial anomalies led the astronomer Nicolaus Copernicus to the realization that the Earth goes around the Sun—not the other way around. In the 1770s, the chemists Antoine Lavoisier and Joseph Priestley inferred the existence of oxygen through experimental results that defied all the theories of the time. Through several decades, plenty of people noticed the strange jigsaw-piece similarity between the east coast of South America and the west coast of Africa, but it wasn't until 1915 that someone pointed out it could be more than a coincidence. Alfred Wegener's insightful observation led to our theory of plate tectonics and continental drift; it is an observation that, at a stroke, did away with the "stamp-collecting" nature of geologic science and gave it a unifying theory that opened up billions of years of Earth's history for inspection. Charles Darwin performed a similar feat for biology with his theory of evolution by natural selection; the days of remarking on the wide variety of life on Earth without being able to tie them all together were suddenly over. It is not just an issue of experiments and observations either; there are intellectual anomalies. The incompatibility of two theories, for example, led Albert Einstein to devise relativity, a revolutionary theory that has forever changed our view of space, time, and the vast reaches of the universe.

Einstein didn't win his Nobel Prize for relativity. It was another anomaly—the strange nature of heat radiation—that brought him science's ultimate accolade. Observations of heat had led Max Planck to suggest that radiation could be considered as existing in lumps, or quanta. For Planck, this quantum theory was little more than a neat mathematical trick, but Einstein used it to show it was much more. Inspired by Planck's work, Einstein proved that light was quantized—and that experiments could reveal each quantum packet of energy. It was this discovery, that the stuff of the universe was built from blocks, that won him the 1922 Nobel Prize for Physics.

Not that a Nobel Prize for Physics is the answer to everything—my view across the Metropole's lobby makes that abundantly clear. Why can't these three men, three of the brightest minds of their generation, see the obvious solution? I can't help wondering if Einstein struggled with that elevator; if he did, by now even he, shaking his fist at the Almighty, would have called out for help.

Admitting that you're stuck doesn't come easy to scientists; they have lost the habit of recognizing it as the first step on a new and exciting path. But once you've done it, and enrolled your colleagues in helping resolve the sticky issue rather than proudly having them ignore it, you can continue with your journey. In science, being stuck can be a sign that you are about to make a great leap forward. The things that don't make sense are, in some ways, the only things that matter.

1

THE MISSING UNIVERSE

We can only account for 4 percent
of the cosmos

The Indian tribes around the sleepy Arizona city of Flagstaff have an interesting take on the human struggle for peace and harmony. According to their traditions, the difficulties and confusions of life have their roots in the arrangement of the stars in the heavens—or rather the lack of it. Those jewels in the sky were meant to help us find a tranquil, contented existence, but when First Woman was using the stars to write the moral laws into the blackness, Coyote ran out of patience and flung them out of her bowl, spattering them across the skies. From Coyote's primal impatience came the mess of constellations in the heavens and the chaos of human existence.

The astronomers who spend their nights gazing at the skies over Flagstaff may find some comfort in this tale. On top of the hill above the city sits a telescope whose observations of the heavens, of the mess of stars and the way they move, have led us into a deep confusion. At the beginning of the twentieth century, starlight passing through the Clark telescope at Flagstaff's Lowell Observatory began a chain of observations that led us to one of the strangest discoveries in science: that most of the universe is missing.

If the future of science depends on identifying the things that don't make sense, the cosmos has a lot to offer. We long to know what the universe is made of, how it really works: in other words, its constituent particles and the forces that guide their interactions. This is the essence of the "final theory" that physicists dream of: a pithy summation of the cosmos and its rules of engagement. Sometimes newspaper, magazine, and TV reports give the impression that we're almost there. But we're not. It is going to be hard to find that final theory until we have dealt with the fact that the majority of the particles and forces it is supposed to describe are entirely unknown to science. We are privileged enough to be living in the golden age of cosmology; we know an enormous amount about how the cosmos came to be, how it evolved into its current state, and yet we don't actually know what most of it *is*. Almost all of the universe is missing: 96 percent, to put a number on it.

The stars we see at the edges of distant galaxies seem to be moving under the guidance of invisible hands that hold the stars in place and stop them from flying off into empty space. According to our best calculations, the substance of those invisible guiding hands—known to scientists as *dark matter*—is nearly a quarter of the total amount of mass in the cosmos. Dark matter is just a name, though. We don't have a clue what it is.

And then there is the *dark energy*. When Albert Einstein showed that mass and energy were like two sides of the same coin, that one could be converted into the other using the recipe $E = mc^2$, he unwittingly laid the foundations for what is now widely regarded as the most embarrassing problem in physics. Dark energy is scientists' name for the ghostly essence that is making the fabric of the universe expand ever faster, creating ever more empty space between galaxies. Use Einstein's equation for converting energy to mass, and you'll discover that dark energy is actually 70 percent of the mass (after Einstein, we should really call it mass-energy) in the cosmos. No one knows where this energy comes from, what it is, whether it will keep on accelerating the universe's expansion forever, or whether it will run out of steam eventually. When it comes to the major constituents of the universe, it seems no one knows anything much. The familiar world of atoms—the stuff that makes us up—accounts for only a tiny fraction of the mass and energy in the universe. The rest is a puzzle that has yet to be solved.

HOW did we get here? Via one man's obsession with life on Mars. In 1894 Percival Lowell, a wealthy Massachusetts industrialist, had become fixated on the idea that there was an alien civilization on the red planet. Despite merciless mocking from many astronomers of the time, Lowell decided to search for irrefutable astronomical evidence in support of his conviction. He sent a scout to various locations around the United States; in the end, it was decided that the clear Arizona skies above Flagstaff were perfect for the task. After a couple of years of observing with small telescopes, Lowell bought a huge (for the time) 24-inch refractor from a Boston manufacturer and had it shipped to Flagstaff along the Santa Fe railroad.

Thus began the era of big astronomy. The Clark telescope cost Lowell twenty thousand dollars and is housed in a magnificent pine-clad dome on top of Mars Hill, a steep, switchbacked track named in honor of Lowell's great obsession. The telescope has an assured place in history: in the 1960s the Apollo astronauts used it to get their first proper look at their lunar landing sites. And decades earlier an earnest and reserved young man called Vesto Melvin Slipher used it to kick-start modern cosmology.

Slipher was born an Indiana farm boy in 1875. He came to Flagstaff as Percival Lowell's assistant in 1901, just after receiving his degree in mechanics and astronomy. Lowell took Slipher on for a short, fixed term; he employed Slipher reluctantly, as a grudging favor to one of his old professors. It didn't work out quite as Lowell planned, however. Slipher left fifty-three years later when he retired from the position of observatory director.

Though sympathetic to his boss's obsession, Slipher was not terribly interested in the hunt for Martian civilization. He was more captivated by the way that inanimate balls of gas and dust—the stars and planets—moved through the universe. One of the biggest puzzles facing astronomers of the time was the enigma of the spiral nebulae. These faint glows in the night sky were thought by some to be vast aggregations of stars—"Island Universes," as the philosopher Immanuel Kant had described them. Others believed them to be simply distant planetary systems. It is almost ironic that, in resolving this question, Slipher's research led us to worry about what we can't see, rather than what we can.

IN 1917, when Albert Einstein was putting the finishing touches to his description of how the universe behaves, he needed to know one experimental fact to pull it all together. The question he asked of the world's astronomers was this: Is the universe expanding, contracting, or holding steady?

Einstein's equations described how the shape of space-time (the dimensions of space and time that together make the fabric of the universe) would develop depending on the mass and energy held within it. Originally, the equations made the universe either expand or contract under the influence of gravity. If the universe was holding steady, he would have to put something else in there: an *antigravity* term that could push where gravity exerted a pull. He wasn't keen to do so; while it made sense for mass and energy to exert a gravitational pull, there was no obvious reason why any antigravity should exist.

Unfortunately for Einstein, there was consensus among astronomers of the time that the universe was holding steady. So, with a heavy heart, he added in the antigravity term to stop his universe expanding or contracting. It was known as the *cosmological constant* (because it affected things over cosmological distances, but not on the everyday scale of phenomena within our solar system), and it was introduced with profuse apologies. This constant, Einstein said, was "not justified by our actual knowledge of gravitation." It was only there to make the equations fit with the data. What a shame, then, that nobody had been paying attention to Vesto Slipher's results.

Slipher had been using the Clark telescope to measure whether the nebulae were moving relative to Earth. For this he used a spectrograph, an instrument that splits the light from telescopes into its constituent colors. Looking at the light from the spiral nebulae, Slipher realized that the various colors in the light would change depending on whether a nebula was moving toward or away from Earth. Color is our way of interpreting the frequency of—that is, the number of waves per second in—radiation. When we see a rainbow, what we see is radiation of varying frequencies. The violet light is a relatively high-frequency radiation, the red is a lower frequency; everything else is somewhere in between.

Add motion to that, though, and you have what is known as the *Doppler effect*: the frequency of the radiation seems to change, just as the frequency (or pitch) of an ambulance siren seems to change as it speeds past us on the street. If a rainbow was moving toward you very fast, all the colors would be shifted toward the blue end of the spectrum; the number of waves reaching you every second would get a boost from the motion of the rainbow's approach. This is called a blueshift. If the rainbow was racing away from you, the number of incoming waves per second would be reduced and the frequency of radiation would shift downward toward the red end of the spectrum: a redshift.

It is the same for light coming from distant nebulae. If a nebula were moving toward Slipher's telescope, its light would be blueshifted. Nebulae that were speeding away from Earth would be redshifted. The magnitude of the frequency change gives the speed.

By 1912 Slipher had completed four spectrographs. Three were redshifted, and one—Andromeda—was blueshifted. In the next two years Slipher measured the motions of twelve more galaxies. All but one of these was redshifted. It was a stunning set of results, so stunning, in fact, that when he presented them at the August 1914 meeting of the American Astronomical Society, he received a standing ovation.

Slipher is one of the unsung heroes of astronomy. According to his National Academy of Sciences biography, he "probably made more fundamental discoveries than any other twentieth century observational astronomer." Yet, for all his contributions, he got little more than recognition on two maps: one of the moon, and one of Mars. Out there, beyond the sky, two craters bear his name.

The reason for this scant recognition is that Slipher had a habit of not really communicating his discoveries. Sometimes he would write a terse paper disseminating his findings; at other times he would put them in letters to other astronomers. According to his biography, Slipher was a "reserved, reticent, cautious man who shunned the public eye and rarely even attended astronomical meetings." The appearance in August 1914 was an anomaly, it seems. But it was one that set an English astronomer called Edwin Powell Hubble on the path to fame.

The Cambridge University cosmologist Stephen Hawking makes a wry observation in his book *The Universe in a Nutshell*. Comparing the chronology of Slipher's and Hubble's careers, and noting how Hubble is credited with the discovery, in 1929, that the universe is expanding, Hawking makes a pointed reference to the first time Slipher publicly discussed his results. When the audience stood to applaud Slipher's discoveries at that American Astronomical Society meeting of August 1914, Hawking notes, "Hubble heard the presentation."

By 1917, when Einstein was petitioning astronomers for their view of the universe, Slipher's spectrographic observations had shown that, of twenty-five nebulae, twenty-one were hurtling away from Earth, with just four getting closer. They were all moving at startling speeds—on average, at more than 2 million kilometers per hour. It was a shock because most of the stars in the sky were doing no such thing; at the time, the Milky Way was thought to be the whole universe, and the stars were almost static relative to Earth. Slipher changed that, blowing our universe apart. The nebulae, he suggested, are "stellar systems seen at great distances." Slipher had quietly discovered that space was dotted with myriad galaxies that were heading off into the distance.

When these velocity measurements were published in the *Proceedings of the American Philosophical Society*, no one made much of them, and Slipher certainly wouldn't be so vulgar as to seek attention for his work. Hubble, though, had obviously not forgotten about it. He asked Slipher for the data so as to include them in a book on relativity, and, in 1922, Slipher sent him a table of nebular velocities. By 1929 Hubble had pulled Slipher's observations together with those of a few other astronomers (and his own) and come to a remarkable conclusion.

If you take the galaxies moving away from Earth, and plot their speeds against their distance from Earth, you find that the farther away a galaxy is, the faster it is moving. If one receding galaxy is twice as far from Earth as another, it will be moving twice as fast. If it is three times more distant, its speed is three times greater. To Hubble, there was only one possible explanation. The galaxies were like paper dots stuck onto a balloon; blow it up, and the dots don't grow, but they do move apart. The very space in between the galaxies was growing. Hubble had discovered that the universe is expanding.

It was a heady time. With this expansion, the idea of a big bang, first suggested in the 1920s, bubbled to the surface of cosmology. If the universe was expanding, it must once have been smaller and denser; astronomers began to wonder if this was the state in which the cosmos had begun. Vesto Slipher's work had led to the first evidence of our ultimate origins. The same evidence would eventually bring us the revelation that most of our universe is a mystery.

TO understand how we know a significant chunk of the cosmos is missing, tie a weight to a long piece of string. Let the string out, and swing the weight around in a circle. At the end of a long string, the weight moves pretty slowly—you can watch it without getting dizzy. Now pull the string in, so the weight is doing tiny orbits of your head. To keep it spinning around in the air, rather than falling down and strangling you, you have to keep it moving much faster—so fast you can hardly see it.

The same principle is at work in the motions of the planets. The Earth, in its position close to the Sun, moves much faster in its orbit than Neptune, which is farther out. The reason is simple: it's about balancing forces. The gravitational pull of the Sun is stronger at Earth's radial distance out from the Sun than at Neptune's. Something with Earth's mass has to be moving relatively fast to maintain its orbit. For Neptune to hold its orbit, with less pull from the distant Sun, it goes slower to keep in equilibrium. If it moved at the same speed as Earth, it would fly off and out of our solar system.

Any orbiting system ought to follow this rule: balancing a gravitational pull and the centrifugal forces means that, the farther something is from whatever is holding it in orbit, the slower it will move. And, in 1933, that is exactly what a Swiss astronomer called Fritz Zwicky didn't see.

As construction began on the Golden Gate Bridge and a forty-three-year-old Adolf Hitler was appointed chancellor of Germany, Zwicky noticed something odd about the Coma cluster of galaxies. Roughly speaking, stars emit a certain amount of light per kilo, so, looking at the amount of light coming out of the Coma cluster, Zwicky could estimate how much stuff it contained. Zwicky's problem was that the stars on the edges of the galaxies were moving far too fast to be constrained by the gravitational pull of that amount of material. According to his calculations, the only explanation was that there was about four hundred times more mass in the Coma cluster than could be accounted for by the cluster's visible matter.

It should have been enough to launch the dark matter hunt, but it wasn't—for the worst of scientific reasons. Comb the Internet for references to Zwicky, and you'll find *brilliant* next to *maverick, genius* next to *insufferable*. Like Slipher, he doesn't figure large in the astronomy textbooks, despite his many important discoveries. He was the first to see that galaxies form clusters. He coined the term *supernova*. He was certainly one of a kind. He built a ski ramp next to the Mount Wilson Observatory in the San Gabriel Mountains of California, for example; in the winter Zwicky would haul his skis to work so he could keep his ski-jumping skills honed. But it was his interpersonal skills that needed most attention. He was a prickly, difficult man, convinced of his own genius, and convinced that he never got the recognition he deserved. He had a tendency to refer to all his colleagues as "spherical bastards": bastards whichever way you looked at them. Small wonder, then, that his colleagues turned a blind eye to his discovery of the Coma cluster's missing mass.

But he was right. Something about the mass of galaxies just doesn't add up—unless, that is, the universe is heavily sprinkled with dark matter. In 1939, at the dedication of the McDonald Observatory in Texas, the Dutch astronomer Jan Oort added to the evidence. Oort gave a lecture in which he showed the distribution of the mass in a certain elliptical galaxy had to be very different from the distribution of the light. He published the data three years later, making this very point clear in the abstract. Again, in a classic Kuhnian response, no one reacted. This spectacular ability to ignore such anomalous results continued for decades until, for some reason, people finally listened to Vera Rubin.

Rubin, who is now in her late seventies, made her first big mark on cosmology at the age of twenty-two. The New Year's Eve, 1950, edition of the *Washington Post* reported on a talk she gave at the American Astronomical Society, hailing her achievements under the headline "Young Mother Figures Center of Creation by Star Motions." The accompanying piece described how Rubin's work was "so daring . . . that most astronomers think her theories are not yet possible." But her most daring work, the fight to get dark matter taken seriously, was still to come.

Not that she even took herself seriously to start with. The story, she says, is a lesson in how dumb a scientist can be. In 1962 Rubin was teaching at Georgetown University in Washington, D.C. Most of her students were from the U.S. Naval Observatory down the road, and they were very good astronomers, she recalls. Together they were able to map out the *rotation curve* of a galaxy. This is a graph that shows how the velocity of the stars changes as you move out from the center of the galaxy. As with that weighted string twirling around your head, the velocities should fall as you get farther out. For Rubin and her naval researchers, though, they didn't; once they got away from the center, the curve was flat. They presented the results in a series of three papers, and Rubin made nothing of it.

Three years later, in 1965, she took a job at the Carnegie Institution of Washington. After a year in the cutthroat business of looking for quasars, the most distant objects known, she wanted to do something a little less competitive, something she could make her own. She decided to look at the outside of galaxies because no one had studied them—everyone concentrated on the centers. Not only had Rubin completely forgotten about her work with the Naval Observatory students, she also didn't believe her own results as she was gathering them. She measured the speeds by looking at how the motion had changed the spectrum of light coming from a star. Rubin was gathering about four spectra each night, gradually going farther and farther out from the center of the galaxy. Even though she developed the spectra as she went along, and they all looked the same, the penny didn't drop.

"You always thought the next point would fall," she says. "And it just didn't."

Eventually, though, she got it. By 1970 Rubin had mapped out the rotation curve for Andromeda; the star velocities remained the same however far out she looked. With the velocities of the stars remaining high at the edge, centrifugal forces should be throwing Andromeda's outer stars off into deep space. By rights, Andromeda should be falling apart. Unless, that is, it is surrounded by a halo of dark matter.

NO one knows what the dark matter actually is. When the Cambridge professor Malcolm Longair wrote his cosmology primer *Our Evolving Universe*, he listed some of the things it might turn out to be. At the top of the list were things like interstellar planets and low mass stars. Toward the bottom of the list were house bricks and copies of the *Astrophysical Journal*. This last candidate seems most appropriate; if it were discovered to be the answer, it would add a pleasing irony to the dark matter story. The *Astrophysical Journal* is where, in 1970, Rubin published her results and brought dark matter in from the cold.

Not that you'd necessarily get that from the paper. The title seems innocuous: "Rotation of the Andromeda Nebula from a Spectroscopic Survey of Emission Regions." The abstract, the summary of the paper, seems to say nothing controversial. The conclusions of the paper are similarly disappointing. It presents the data—measurements of the rotation speeds of the stars in Andromeda—and says nothing more. The graph from page 12 is still on the wall of Rubin's office at the Carnegie Institution's Department of Terrestrial Magnetism in Washington, D.C., however. And today it remains just as relevant, and just as mysterious, as it was on publication.

The idea of a clutch of invisible matter holding on to Andromeda's outer stars didn't catch on straightaway, but at least this time it wasn't ignored. First, astronomers justified the blind eye they had turned for thirty-seven years. They started constructing their own rotation curves, for example, by coming up with exotic explanations for how the mass might be distributed through the galaxies. None of these efforts ever convinced Rubin, she says; somehow, a couple of the points were always so far off the curve—and ignored—as to make the ideas laughable.

By the 1980s astronomers had given up trying to fudge the data. Something about the gravitation of galaxies didn't fit, and the best explanation was the existence of some matter that didn't shine like the stars, or reflect light, or give off detectable radiation, or behave in any way that would make its presence known—except by its gravitational pull. The quest was now on to find out what this strange stuff was.

The first meeting on the subject of the new dark matter was held at Harvard University in 1980. Rubin then confidently proclaimed to the audience that we would know what dark matter was in just a decade. That deadline came and went, and we were none the wiser. In 1990, at a meeting in Washington, D.C., Martin Rees, the English astronomer royal, told an audience that the mystery would be solved within ten years. Then, in 1999, one year away from the deadline he had imposed in Washington, Rees gave an extension, declaring, "[I am] optimistic that if I were writing in five years' time, I would be able to report what the dark matter is."

His optimism was misplaced. We still don't know what the dark matter is. A series of exotica have been suggested, everything from black holes to as-yet-undiscovered particles with extraordinary properties. Nothing that fits the bill has yet been discovered. And it's not for want of looking.

SEARCHING for dark matter is not for the fainthearted; the stuff has eluded detection for thirty years for good reason. Nevertheless, scientists do have some ideas of how to look. Physicists have models for what kind of particles might have been created in the big bang that could still be hanging around in the cosmos to act as dark matter. Their best guess is something called *weakly interacting massive particles,* or *WIMPs.* If this is right, there's no shortage of dark matter to hunt for. According to the particle physicists, the Earth is drifting through a mist of dark matter right now; something like a billion WIMPs are washing through your head every second.

Among the WIMPs, there is one outstanding candidate: the neutralino. It is stable enough to still be filling the cosmos 13 billion years after the big bang. It would be suitably difficult to see or feel; it doesn't interact via the strong force that holds nuclei together, and it ignores and is ignored by electromagnetic fields. Crucially, it has enough mass—about one hundred times the mass of a proton—to have the necessary effect on galaxies. The only drawback is that no one knows whether the neutralino really exists.

If you want to find experimental evidence for dark matter, you have to get it to interact with something. Our best chance of that comes with atoms that have large nuclei. The dark matter hunters use large arrays of silicon or germanium crystals, or huge vats of liquid xenon. The hope is that one of the WIMPs will make a direct hit on one of these fat atomic nuclei. If that happens, the nucleus should recoil a little bit (in the case of the crystals) or send out an electrical signal (from the liquid xenon). There are a couple of complications, though.

First, the nuclei vibrate naturally anyway, so physicists need to hold them still in order to avoid a false detection in the apparatus. The crystal arrays, for instance, have to be cooled down to a fraction of a degree above absolute zero, the temperature where everything stops moving. Cooling the detectors this much is cumbersome and difficult. And then there's the second complication: cosmic rays.

Earth is continually bombarded by high-speed particles from space, and these cosmic rays produce exactly the same signature as WIMPs in a WIMP detector. So the searches have to take place deep underground, beyond the reach of the rays. It is a complication that makes the dark matter hunters the inhabitants of some of the most inaccessible laboratories on Earth. An Italian group have put their detector under a mountain. The neutralino search in the United Kingdom takes place 1,100 meters underground, in a potash mine whose tunnels reach out under the seafloor. U.S. researchers have set up a dark matter hunt seven hundred meters underground, in an abandoned iron mine in northern Minnesota.

When you understand the working conditions, you know these people must be serious. And yet, so far, they have found precisely nothing. The searches have been going on for more than a decade; indeed, many of the researchers have dedicated more than two decades of their lives to the quest for dark matter. Upgrades are making the equipment more sensitive all the time, but we still have no defensible idea of what is causing that strange pull in the heavens.

It seems somehow impossible that, when this stuff makes up a quarter of the universe, we don't know yet what it is. But we should perhaps take comfort in the fact that we at least noticed it was missing. If we hadn't, it's hard to imagine how wrong we'd have got things when, in 1997, it became apparent that another bit of the universe was also absent without leave. If dark matter was a problem, the discovery of dark energy was a catastrophe.

IF the universe is expanding, as Hubble showed it is, two questions spring immediately to mind. First, how fast is it expanding? Second, will it keep expanding forever?

The answer to the first question comes from measuring the velocities of the receding galaxies, and knowing how far away they are. You can't just measure how fast a galaxy is moving away from us and call that the expansion rate of the universe; the way space expands messes with your common sense. The farther away from us a galaxy lies, the faster it is moving away from us because the space in between Earth and the galaxy is also expanding. The result, known as *Hubble's constant*, gives a measure of the expansion rate; currently, we think it is about seventy kilometers per second per (roughly) 3 million light-years. The accuracy shouldn't be taken too seriously; that value is always subject to change when a better set of measurements come in.

Answering the second question is, in many ways, much more interesting. If the universe is still expanding after the big bang, that expansion should be slowing down; the mutual pull of all the matter in the universe works against any further expansion. So our cosmic future depends on how much stuff there is out there, and how it is arranged.

Cosmologists already know something about those questions from one very easy scientific observation: the fact that we exist. For that to be the case, the universe must have expanded from its hot, dense beginnings with a particular amount of energy. If there had been too much, any matter that was created would have been spread so thinly that gravity couldn't have pulled atoms together into stars, galaxies, and—eventually—humans. As the matter spread farther, its gravitational pull would have become even weaker and the expansion energy ever more dominant. The universe would have blown itself apart before anything interesting—humans, for example—happened.

If there had been too little expansion energy, on the other hand, gravity would have pulled all the matter together in a similar feedback cycle: once things got closer together, their gravitational pull would have become stronger, pulling them even more. Eventually, the fabric of the universe would have shrunk back to implode in a scenario astronomers call the *big crunch*.

Given a certain amount of expansion energy, producing a Goldilocks universe like ours—one that's "just right"—involves a precise distribution of matter. As a shorthand for talking about the density of gravitating matter, astronomers refer to the *Omega* value of the universe. An Omega of 1, which corresponds to a measly six hydrogen atoms per cubic meter of universe (a cubic meter of the air around you has something like 10 million billion billion atoms), is where the matter density more or less balances out the expansion.

According to theory, the existence of stars and galaxies relies on Omega starting out within one part in a million billion of 1. And, because of the nature of the feedback cycle with Omega, starting out in balance means remaining in balance. Today, if the theorists are right, Omega should still be near 1. The trouble is, we know that there's not nearly enough matter—dark or otherwise—to make Omega 1.

It is this problem that led to the return of Einstein's cosmological constant, something that no one saw coming. Hubble's triumphant discovery of the universe's expansion had meant the cosmological constant could be ditched. The equations of general relativity simply didn't need the fudge factor that produced a steady-state universe, and by 1930 Einstein's antigravity lay embarrassingly redundant. Who could have imagined that, nearly seventy years later, it would be back, reincarnated in the ghostly form of dark energy?

ASTRONOMERS first started investigating Omega in the 1930s as a means of predicting the fate of the universe. If Omega is indeed 1, the expansion will continue at its present rate. If the theorists are wrong, and Omega is less than 1, the power behind the expansion will increase as the matter thins out. If Omega turns out to be greater than 1, gravity will eventually win out, and our future lies in a big crunch.

Initially, the astronomers investigated Omega by continuing Slipher and Hubble's methods: measuring the properties of the light from galaxies. The vast number of light sources in a galaxy meant that this never produced anything reliable, however; it is rather like trying to measure the properties of human speech by listening to the noise of a soccer crowd. What they needed was a single object, something whose properties you could measure and draw inferences from. In 1987 they found one. If you want to understand the fate of the universe, it turns out you're going to have to get to grips with exploding stars: supernovae.

We've been seeing supernovae in the skies for centuries; the Danish astronomer Tycho Brahe reported seeing one in 1572, more than thirty years before the invention of the telescope. They occur when a star gets too big and collapses under its own gravity. During the few weeks or months over which this collapse takes place, transforming the star into a neutron star or even a black hole, it shines with the power of 10 billion suns. On Monday, February 23, 1987, we saw such a sight. The explosion of Sanduleak-69 202, a blue giant star in the Large Magellanic Cloud galaxy, was notable for two reasons. First, because it turned the star into the brightest supernova seen since 1604. Second, because its light was the first to give a standard for measuring cosmic distances.

The way certain supernovae—they are known as Type 1a Supernovae—emit their light has a peculiar characteristic that makes them supremely appealing to astronomers. Type 1a explode because they have sucked too much material from a nearby star. Analyze the spectrum of the light from this kind of explosion, and how fast its brightness fades away, and it will tell you how far the light traveled to Earth, and how the expansion of space stretched the light on its journey.

The only drawback is that you have a limited window of opportunity. With supernovae, timing is everything. If you want to get useful information, you have to find it within a couple of weeks of the light first reaching Earth. Since an explosion happens about once per century in each galaxy, that means scanning a lot of galaxies with your telescope.

This kind of drudge work is a long-standing problem for astronomers. Inside Flagstaff's Lowell Observatory, for example, you can experience the agonizing nature of astronomy in Slipher's day. When he led the search for Pluto, the technique used was a celestial Spot-the-Difference. Put two photographic plates of the same region of the sky, taken on different nights, into a machine called a blink comparator, and you can shuttle between the two almost entirely similar views. The winner is the first to spot the one white dot—in the mess of white dots—that has moved. That shifting white dot is the planet you are looking for.

Fortunately, in the Lowell exhibition, someone marked the displaced dot with a big white arrow. Modern image-reading technology has made spotting the appearance of a supernova even easier; today, we have computers to provide the big white arrow. They can compare two different photographs of the sky, then highlight the differences. Some of those will be asteroids; some will be the varying brightness associated with the black holes at the center of galaxies; some will be false signals—bright flashes from subatomic particles hitting Earth's atmosphere. And, just occasionally, one will be a distant supernova.

The first strong interpretations of supernova data came in June 1996 from a team based at California's Lawrence Berkeley National Laboratory (LBNL). This announcement was made at a cosmology meeting convened to celebrate the 250th birthday of Princeton University, the adopted intellectual home of Albert Einstein. A perfect place to begin the resurrection of his cosmological constant, as it turned out.

When astronomers first got close to using supernovae to chart the universe's expansion, they were convinced they were going to find a deceleration. After all, the power of the big bang should be running out; gravity had taken over, and the brakes were firmly on. It turns out, though, that the universe is not so simple.

At first glance, the LBNL results confirmed suspicions. The supernova light suggested that the universe's expansion was slowing down: the gravitational pull of the universe's contents was decelerating the cosmos and setting Omega to somewhere around 1.

But it was a controversial finding. All the known gravitating matter in the universe—including the dark matter—gave an Omega of only 0.3. Had everyone underestimated the amount of dark matter? It seemed unlikely; by this time various different methods for determining the mass of galaxies were in use, and each showed there was significantly more gravitating matter than we could see, and each gave approximately the same numbers.

If dark matter was on a fairly solid footing, what was going on? The cosmologists Michael Turner and Lawrence Krauss were at the Princeton meeting, and they had an answer ready. Why not keep the dark matter at 0.3 but let something else make up the missing 0.7? Instead of looking for some extra matter, why not assume it is actually extra energy? Bring back Einstein's cosmological constant, they said.

As is proper, experiment won out over the theorists' speculations. When Saul Perlmutter published his LBNL group's results, the supernova data indicated that gravitating matter could account for pretty much all of Omega. No one needed to bring back the cosmological constant; someone just needed to sort out the dark matter discrepancy. There must be more out there.

The trouble was, Perlmutter's results raised problems of their own. If you know the matter density in the universe, the current expansion rate (Hubble's constant), and how much the universe's expansion is slowing down, you can work out how long it is since the expansion started; the age of the universe, in other words. With an Omega of 1 that is entirely due to matter, the deceleration from the Lawrence Berkeley data put the universe's age at not more than 8 billion years old. Unfortunately, astronomers who had analyzed the light from the universe's oldest stars set *their* age at around 15 billion years old. It doesn't take a Harvard-trained mind to work out that the universe simply can't be 8 billion years old if the stars are nearly twice that age. If there was a problem with the cosmological constant making up Omega, there was also a problem with having a matter-induced Omega of 1. The only reliable fact, it seemed, was that dark matter made up 0.3 of Omega; everything else was up for grabs.

Not everyone was disappointed by this impasse; Robert Kirshner, for one, was rather pleased. The Harvard astronomer was worried that his own supernova results were coming too slowly to compete with the LBNL team; that his team had been beaten to the punch. But it seemed the race to understand the fate of the universe was still wide-open.

In his book *The Extravagant Universe* Kirshner tells the story behind the supernova searches and the reinstatement of Einstein's cosmological constant with great clarity and wit. In the end, he turned the tables and came out first with the result that defined a new era in cosmology. But only after he had defeated his own prejudices.

Kirshner's team, composed of a handful of researchers from all over the world, was using supernova observations from telescopes on mountaintops in Chile, Arizona, and Hawaii. Like the LBNL group, they would look out for new supernovae, month after month, then follow up any really promising candidates by taking over the Hubble Space Telescope for some detailed observations. Hubble could tease out information on a supernova's distance from Earth and how the spectrum of its light varied as the explosion ran its course.

Eventually, they had what they needed. And they didn't like it one bit.

The distant explosions were fainter than they should have been: the light was having to travel farther than it should have. It was Adam Riess, a Berkeley-based astronomer on Kirshner's team, who first said it out loud: the data pointed to an acceleration. The expansion of the universe was speeding up.

It was impossible. But try telling that to the supernovae. Every time Riess used the supernova data—the luminosity, the redshift, and the fade over time—to work out a value for Omega, his calculations told him the universe contained a negative amount of mass. The only way to make sense of it was to assume that mass wasn't the only force at work in the universe's expansion. Add in a cosmological constant, and it all made sense. Given the choice between invoking negative mass and resurrecting Einstein's long-abandoned cosmological constant, the constant won out. But only just.

By January 1998, it was clear from conference presentations that the LBNL team's data were now also pointing in the same direction; they had refined their analysis, sorting out problems like how to account for the way interstellar dust would affect the observations. The thing was, no one wanted to get it wrong. Announcing the return of Einstein's cosmological constant became a battle of nerves, a test of each team's faith in their experimental abilities. To make the claim, or to wait a little, do a few more tests, look again for the mistakes in handling the data? The prize was to be first to produce the scientific result of the decade. The risk was sharing Einstein's egg on the face.

Kirshner didn't like the result, and he certainly didn't want to taste any egg. He admits to doing everything he could to make this go away. On January 12, 1998, he e-mailed Riess with some advice. "In your heart, you know that this is wrong," he wrote.

Riess replied that evening in a long e-mail to the team. His reply sounds almost Shakespearean, like something Henry V might have said if he was an astrophysicist. "Approach these results not with your heart or head but with your eyes," he wrote. "We are observers after all!"

At the end of February, they came out with the results. A media storm followed. Riess eloquently told the CNN audience that the universe's expansion was accelerating, the cosmos was literally blowing itself apart—and Einstein's cosmological constant was back, pushing on the fabric of the universe. Kirshner came out with a rather un-Shakespearean sound bite, reported on February 27 in the *Washington Post*. "This is nutty-sounding," he admitted. "But it's the simplest explanation."

Not that they were happy about it even then. The team leader, Brian Schmidt, probably put it best. His reaction, he told *Science* magazine, was "somewhere between amazement and horror."

Nevertheless, the LBNL came out with the same conclusions shortly afterward. The results still stand. And what is pulling the universe apart? We simply don't know. But it is also pulling at the threads of the ultimate quest in physics.

BRIAN Schmidt's amazement and horror cannot have begun to plumb the depths of amazement and horror that would follow from his team's announcement. This is no longer just a cosmological mystery. The "nutty-sounding" observation, based on the light emitted by a series of exploding stars, created rifts between some of the most eminent scientists on the planet. Now that the cosmological constant is back in play, no one can agree how best to proceed. Paul Steinhardt, a theorist at Princeton University in New Jersey, expressed his dismay that, thanks to the "cosmological constant problem," many of our finest minds seem to have given up on ever understanding our universe. "I'm disappointed with what most theorists are willing to accept," he told the journal *Nature* in July 2007.

The controversy is—quite literally—much ado about nothing. The nothing in question is the universe's "empty" space, which is, in reality, far from empty.

The cosmos, whether it contains any matter or not, is fizzing with energy. In the 1920s, shortly after the birth of quantum theory, which describes how nature behaves at the scale of atoms and subatomic particles, the British physicist Paul Dirac used it to produce a quantum version of the theory behind the characteristics of electric and magnetic fields. Dirac's *quantum field theory* eventually led to the prediction that empty space has energy. Since physicists refer to empty space as *the vacuum*, Dirac's energy has come to be known as the *vacuum energy*.

According to our best guess, this vacuum energy must be what powers the "antigravity" acceleration uncovered by the supernovae; the vacuum energy is the cosmological constant. The trouble is, the measurements from the supernovae tell us the vacuum energy is tiny. It is usually measured in grams. (Remember, according to Einstein's famous equation $E = mc^2$, mass and energy are interconvertible.) The amount of vacuum displaced by the Earth's volume in space would contain about one hundredth of a gram's worth of vacuum energy. That's how small it is.

When, however, theorists work out the vacuum energy from quantum field theory, they get a number that is too big. Massively too big. Their theory suggests that the vacuum energy is so big, it should have ripped the universe apart already in one massive hyperacceleration. This is known as the cosmological constant problem and is widely accepted—even by the physicists involved—as *the* most embarrassing mismatch between theory and experiment ever. A million is a big number: a 1 followed by 6 zeroes. A trillion has 12 zeroes. The mismatch between the measured and the theoretical value for the cosmological constant has 120 zeroes. One hundred and twenty.

Faced with this failure, many physicists have adopted an idea first raised by the Nobel laureate Steven Weinberg in 1987. In his book *Dreams of a Final Theory*, Weinberg suggested that a cosmological constant might exist in our universe without us ever being able to explain its value. If ours was just one universe among many, each might have different values for its constants. Some of these universes would no doubt be sterile, but some would lead to the production of life; there would probably be at least one where things like humans evolved. This is the *anthropic landscape* approach to explaining the nature of the universe. (*Anthropic* means "of humans.") The approach, when you boil it down, essentially says that our universe is the way it is because otherwise we couldn't be here to observe it. It doesn't necessarily invoke a designer or any intention; it simply means if conditions were different, no one would be around to observe them. Essentially, it says the very fact that we observe the universe limits the range of forms it can take. The landscape bit comes from the physicists' assertion that our universe is composed of a hugely varied terrain, a patchwork quilt of subuniverses, each with its own unique and randomly assigned properties. There need be no explanation for the values of the constants in each one.

As an "explanation" for the value of the cosmological constant, this is, to many physicists, abhorrent. Weinberg's suggestion is, says the Stanford University physicist Leonard Susskind, "unthinkable, possibly the most shocking admission that a modern scientist could make."

The idea is so distasteful because it turns science on its head. The philosopher Karl Popper said that science progresses only by falsification: Someone throws up a hypothesis, and then anyone can use experimental data to attempt to shoot it down. If the data falsify the hypothesis, you move on to the next one. Only when you have a hypothesis that has survived many shots can you start to place some faith in what it's saying.

With the anthropic landscape, this approach doesn't work because the other universes are out of reach. You can't falsify the notion because you can never test it with experimental data. No longer do we explain why the universe is as it is; instead, the universe is as it is because that makes it the kind of universe we can inhabit. Is this science? It might just be, Susskind says; he thinks Weinberg is probably right. If we are to make progress toward understanding the universe, we may now have to ditch Karl Popper and his adherents—Susskind calls them the *Popperazzi*—as the ultimate arbiters of what science is and isn't. Perhaps we should just accept that, however much it makes the Popperazzi fume, the laws of our universe may be as they are because of our own existence.

Difficult as this notion is to swallow, there is reason to take it seriously. Quantum field theory suggests that, if we must use a cosmological constant to complete our description of the universe, our universe really ought to be one of very many. It may be that, as E. E. Cummings once wrote, "there's a hell of a good universe next door."

At the root of this argument is the *uncertainty principle* of quantum theory, which says the fundamental properties of any system are never exactly defined but have an intrinsic fuzziness. The uncertainty principle, when applied to quantum field theory, produces natural fluctuations in the properties of certain regions of the universe. It is rather like having a balloon that is peppered with weak spots; as the universe inflates, these fluctuations can grow, producing a new region of space and time. In other words, a universe containing a cosmological constant that arises from the vacuum energy will produce new bubble universes all the time. Those bubbles will produce their own new baby universes in turn—and so on, ad infinitum. What we think of as the universe is only one region of space-time in a frothing sea of mini-universes.

The anthropic landscape idea has many supporters now, especially among theorists; that is why Steinhardt puts himself in the minority. But if we can't access these bubble universes to see whether they have different constants, aren't we effectively giving up on physics?

This was the root of the discussion in Brussels, the ghost of Albert Einstein looking over every shoulder. Should we be shrugging our shoulders and putting the value of the cosmological constant down to the particular kind of universe we live in? Can we face the idea that we may never understand what most of the universe is, that we may never get to the root of dark energy? The answer was both yes and no: yes, it is a possibility we have to face; no, it doesn't mean giving up hope of an explanation. David Gross, who chaired the conference, was quick to make the point that at the first Solvay conference in 1911, the physicists were similarly puzzled. Some materials had been shown to be emitting particles and radiation in a way that seemed to violate the laws of conservation of mass and energy. The explanation came a few years later, when quantum theory was developed. "They were missing something absolutely fundamental," Gross told the 2005 Solvay assembly. "We are missing perhaps something as profound as they were back then."

So what is that "something fundamental"? Do we have any clues? The answer depends on whom you ask. Adam Riess, the man whose radical, Shakespearean rhetoric pulled us into the dark energy era, offers a provocative suggestion. What if, he says, we just don't know enough about how gravity works? Maybe there isn't any dark matter, and maybe there isn't any dark energy. Maybe for the last four centuries we've all been blind to tiny inaccuracies in Newton's law of gravity, and these inaccuracies hold the key to restoring the lost universe.

Riess isn't the first to raise the idea, and he's not saying it necessarily has any merit. His point is that it is a possibility, and it has yet to be ruled out. Vera Rubin feels the same. She reckons that ninety-nine physicists out of a hundred still believe in the existence of some dark stuff that fills the universe, its gravitational influence holding galaxies together. But, to her eyes, changing the fundamentals of physics is starting to look like a better option.

On the face of it, the fix can be a relatively simple one. It was first suggested in 1981 by an Israeli physicist called Mordehai Milgrom. Basically, you tweak Newton's law of gravity so that at large distances, the kinds of distances that stretch across galaxies and even clusters of galaxies, gravity is a little bit stronger than you'd otherwise expect. The idea is known as *Modified Newtonian Dynamics*, or *MOND*, and—despite its apparently innocuous nature—it has caused a lot of trouble.

Taking something that has worked perfectly well for four hundred years, something that was created by a man widely considered to be the greatest scientist of all time, and suggesting it needs a little tweak is a brave move. Milgrom was not taken seriously when he first suggested it. But he did gain a few supporters. Most notable among them was a young astronomer named Stacy McGaugh.

MCGAUGH has taken so much flak in defense of MOND, he should be issued with a Kevlar jacket. If the way the dark matter problem was overlooked for forty years taught Vera Rubin how dumb scientists could be, McGaugh, who used to be one of her graduate students, taught her something else: just how resistant science is to change.

In March 1999 McGaugh gave a talk on MOND at the Max Planck Institute in Germany. No one there was willing to embrace the idea. If you want us to take you seriously, they said, predict something; when it is borne out by experiment, we'll listen.

A few months later McGaugh published a paper in the *Astrophysical Journal* that asked the cheeky question "What if there is no dark matter?" The result, he said, would be that a characteristic feature in the cosmic microwave background radiation, the echo of the big bang, would be different from what the dark matter advocates expected. The *power spectrum*, a kind of breakdown of the radiation, would show it up. Both MOND and dark matter models predicted that the power spectrum would take the form of a series of peaks and troughs. Dark matter theory said the second peak would be slightly lower than the first, but not significantly. Without dark matter, that second peak would be tiny, McGaugh pointed out; let's see what happens when the data come in.

McGaugh's paper was published in late 1999. In the summer of 2000 Rubin was at a conference in Rome, watching him give a presentation based on his paper to an audience of astronomers. Now there were data. And there was no second peak. None at all.

McGaugh had been granted a ten-minute slot. Rubin watched in shock as, when McGaugh ended his talk, nothing happened. "There was not a single question afterwards," she recalls. What's more, she adds, the next morning some eminent cosmologist started the discussion of the new results with not a single mention of the fact that they were different from the accepted dark matter model.

Rubin has been impressed by MOND from that time on. Partly because she doesn't like the idea of invoking exotic new particles to explain a straightforward observation, and partly because mainstream astronomy has gotten too good at public relations, and good PR, she says, suppresses proper scientific debate. Rubin has always been a fan of the underdog in science.

For a long time, MOND wasn't even an underdog. As McGaugh will testify, it was more like a mangy dog sitting outside the conference hall: an ad hoc idea cobbled together by an Israeli physicist with no better rationale for modifying gravity than the majority had for invoking dark matter. But then, in 2004, Jacob Bekenstein got involved.

Bekenstein was born in Mexico City, studied physics at the Polytechnic Institute of Brooklyn and Princeton University, and is now a professor at the Hebrew University of Jerusalem. As a young man he got up Stephen Hawking's nose by making various controversial proposals about black holes (which all turned out to be correct); now he is simply seen as one of our most formidable minds. As soon as Bekenstein developed a version of Einstein's relativity specifically tailored to show why MOND should be taken seriously, the physics world had no choice but to sit up and listen. When Bekenstein's relativistic MOND started fitting rather nicely with other observations of the galaxies, what had once been a fringe idea suddenly had to be taken seriously. And when lifelong dark matter supporters started switching sides, things started to get ugly.

SOMETIMES, the idea that science is a neutral, careful, bias-avoiding discipline has a bad day. One such day was August 21, 2006, when a NASA press release crowed, "NASA finds direct proof of dark matter."

The crowing was over observations of a massive collision between two clusters of galaxies, known collectively as the Bullet Cluster. Observing the outcome of the collision, astronomers had found that dark matter had separated from normal matter. They inferred this from the way that light bent around a seemingly empty area of space. One of Einstein's great successes was to show that mass and energy distort the very fabric of the universe. Any radiation—be it light or X-rays—traveling through space dotted with massive stars and planets will therefore follow a curved path rather than a straight one. So when NASA's Chandra telescope recorded light bending around empty space, with no visible matter in the vicinity, it seemed like a slam dunk for dark matter and a poke in the eye for the troublemakers who claim there's no need to invoke dark matter, pixie dust, or magic space blancmange (as one satirist decided to call it) to explain the universe.

The press release put the mainstream case majestically. "A universe that's dominated by dark stuff seems preposterous, so we wanted to test whether there were any basic flaws in our thinking," said Doug Clowe of the University of Arizona at Tucson, and leader of the study. "These results are direct proof that dark matter exists."

Except that they're not, exactly. They are, the press release later concedes, simply "the strongest evidence yet that most of the matter in the universe is dark."

The release went on to gasp that some have had the gall to doubt the existence of dark matter. They could no longer, apparently. "Despite considerable evidence for dark matter, some scientists have proposed alternative theories for gravity where it is stronger on intergalactic scales than predicted by Newton and Einstein, removing the need for dark matter. However, such theories cannot explain the observed effects of this collision."

It was all over for modified gravity theories, you'd think. Except it seemed that no one had actually asked the modified gravity people whether their theories could or couldn't explain the observed effects of the collision. In fact, no one had even checked the archive of papers where physicists routinely post their latest results and theories.

Two months before the triumphant NASA announcement, researchers looking at Bekenstein's relativistic MOND theory had taken a glance at the Bullet Cluster. Their paper, playfully titled "Can MOND take a Bullet?" and published in a well-respected peer-reviewed astronomy journal, makes interesting reading. There was nothing in the Chandra observations that contradicted relativistic MOND, it argued. Milgrom's reaction was also intriguing. We heard the same claims three years ago, he said; the MOND community has had plenty of time to digest the matter, to discuss it at conferences, and to let the authors know how MOND explains it, "but they don't seem to listen." In McGaugh's view, the Bullet Cluster is difficult for MOND to explain without invoking some unseen matter, but there's no need for anything exotic. The presence of some neutrinos (which are known to exist, are difficult to detect, and make up some small fraction of the dark matter in the standard theory) might be enough to explain the observations. Plus, McGaugh points out, we know that the kinds of particles we are made of—they are called *baryons*—make up 4 percent of the cosmos, but we've only ever directly detected one tenth of the baryons that are known to exist. Maybe these "dark baryons" are involved in the Bullet Cluster?

MOND, accompanied by neutrinos and dark baryons, wasn't even the only alternative. Nine days after the NASA press conference, the Canadian physicist John Moffat posted his response on the archive. His modified gravity theory, he said, could also explain the Chandra observations without invoking any dark matter.

Moffat is one of those rarest of scientists: he is self-taught, having left Paris as a penniless artist, and yet has risen to occupy senior academic positions. His story reads like a fairy tale: In 1953, at the age of twenty, he wrote a letter to Einstein, expounding on some implications of the great man's ideas. Einstein wrote back, impressed with Moffat's work and understanding, and started to open doors for the young man. By 1958 Moffat had a PhD from Trinity College, Cambridge—without ever earning an undergraduate degree.

Not that luck has always been on Moffat's side. His unconventional genius led him to work on unfashionable ideas, and in science fashion matters. He had his biggest idea—that the speed of light might have been different in the past—around a decade too early. Though Moffat only managed to publish it in an obscure journal in the early 1990s, the idea came to the forefront of physics ten years later. Even then, Moffat had to kick up a fuss before he got any proper recognition.

And he is still kicking up a fuss—but now in the realm of dark matter. Moffat's explanation for the flat rotation curves of galaxies is called, rather inelegantly but at least unpretentiously, *MOG. Modified Gravity*—that's it. But according to Moffat, MOG's slight adjustment to Newtonian gravity, making it a little stronger than normal at large distances, explains the Chandra observations.

Maybe dark matter is there; maybe it is not. There are alternatives, and any neutral observer has to say the dark matter issue has not yet been resolved. So far, we've waited more than sixty years to find out what is causing those strange galactic rotations, and it is possible that none of us alive today will ever find out the truth about dark matter. Maybe we'll know tomorrow. Until we do, though, as Adam Riess pointed out, we can't be sure about dark energy.

NOT that the dark energy researchers are twiddling their thumbs. NASA, the National Science Foundation, and the U.S. Department of Energy have commissioned a group of physicists to find the best way forward for exploring the dark energy enigma, and in September 2006 the Dark Energy Task Force issued their report. Most of their conclusions recommended an "aggressive program" of experiments and astronomical observations that will help us make sense of it all. What is most intriguing, though, is that, besides all the program recommendations, the chair of the task force quietly recommended another way to approach the dark energy issue. What we really need, says Edward "Rocky" Kolb, is another Einstein.

Kolb suggested that dark energy might be solved by winding physics back eighty-five years. Part of the problem, he says, might be the assumptions theorists made in the 1920s in order to find solutions to Einstein's equations (the solutions are, essentially, mathematical descriptions of the universe). They assumed that the universe was *isotropic*, that is, pretty much the same, whichever way you looked at it.

If it's not too peculiar a notion, imagine standing inside a blueberry muffin and looking around. The blueberries surround you left, right, up, and down; whichever way you look, there's no appreciable difference in how they are distributed throughout the muffin. Our view from inside the universe appears to be the same. Sure, if we look one way in the solar system or the Milky Way, we'll see certain familiar features that aren't there if we look the other way. Once we look beyond our local region, however, the universe seems the same wherever we look.

But is it? We don't know for sure. There are rumblings among astronomers that measurements of the cosmic microwave background radiation, the echo of the big bang, are showing hints that the universe is not isotropic and some cosmologists are suggesting there is good reason to consider bringing back a concept dismissed at the end of the nineteenth century: the *ether*, a ghostly entity that makes it easier for light and particles to move through the universe in one direction rather than another. Either scenario would invalidate the assumption of isotropy. At the moment, we don't have enough information to know anything for sure, but it is clear that, to get closer to the truth about the missing universe, what we really need is a theory that doesn't make the assumption. Only with that theory in place can we be sure we haven't led ourselves into error.

It's easier said than done. Put bluntly, we are not yet clever enough to describe the universe without making those—possibly catastrophic—simplifying assumptions. It's not an impossible puzzle, as far as we know. It's just that we stand without the required insight—we can't yet do the math. We are like the generation before Einstein. But one day, Kolb says, someone will work out how to solve Einstein's equations without the crippling assumptions of isotropy, and that person might then throw out something interesting, something like an explanation for dark energy. On that day, the inaccessibility of the landscape of universes—if such a thing exists—would no longer have any bearing on our understanding of the cosmos.

IT'S certainly something to look forward to. For the moment, however, all we can do is be Slipher-conservative and declare with confidence that there is more to the universe than we currently know. The cosmos is still ripe for investigation.

Who knows what surprises it has in store? Especially since dark energy and dark matter are not the only hints that there are things out there waiting to be brought into the canon of physics. There are reasons to doubt, for example, that what we call the laws of physics necessarily apply everywhere in the universe—or that they were applicable to every time in its history. That would surely change our view of the universe's evolution. Before heading off down that trail, though, we should first examine the tale of two spacecraft, launched in the 1970s. They are currently leaving our solar system—but on a very slightly, and mysteriously, different course than the one with which they were programmed. Perhaps the Pioneer anomaly can tell us what's wrong with our cosmos.

THE PIONEER
ANOMALY

Two spacecraft are flouting the
laws of physics

saac Newton offers hope to every underachiever. He was born prema-
turely, a runt among newborns who, according to his mother, could be
"put in a quart mug." At his school he was among the poorest performers.
Then, at the age of twenty-three, he came up with the *universal theory of
gravitation.* There is a force between any two bodies, it said, that is "directly
proportional to the product of their masses and inversely proportional to
the square of the distance between them."

Though it might seem simple, it is, quite literally, rocket science. Everything we launch into space is governed by this inverse square law because
rocket scientists have to apply it to understand how their craft will move
through the gravitational fields of the planets and moons of our solar system and—as in the case of the Pioneer probes—beyond.

By rights, the Pioneer 10 and 11 space probes should no longer be of interest to anyone. Launched in the 1970s, they are now far beyond the edge
of our solar system, drifting silently out into the void. The last contact we
had with Pioneer 10 was on January 10, 2003, when a weak signal made it

back to Earth. It is now nearly 8 billion miles away, past the orbits of Neptune and Pluto, and we will not hear from it again because it no longer has any power left with which to send out a signal. The probe's next significant moment will come in 2 million years, when, according to calculations based on the gravitational law that Newton developed just over three centuries ago, it will hit the star Aldebaran in the constellation Taurus.

However, the Pioneer probes hint that the law might be wrong, or at least wrong for those particular calculations. For the probes are drifting off course. In every year of travel, the probes veer eight thousand miles farther away from their intended trajectory. That is not much when you consider that they cover 219 million miles a year; whatever is causing the drift is around 10 billion times weaker than the Earth's pull on your feet. Nonetheless, it is there, and casting doubt over the universality of one of Newton's greatest achievements.

The idea that the Pioneer probes threaten the known laws of physics is almost universally derided—even by the people trying to make sense of the anomaly. The fact that is seldom appreciated, though, is that NASA explicitly planned to use them as a test of Newton's law. The law failed the test; shouldn't we be taking that failure seriously?

IN 1969, when most eyes were on the Apollo moon landings, John Anderson was focused on the Pioneer probes. As principal investigator, he had the job of making sure they would do everything they should—that is, observe the outer planets. It dawned on Anderson, however, that they could do more.

As spacecraft, the Pioneer probes are unique. Every other craft has the means of checking its orientation and trajectory—by triangulating its position with certain stars, for example. If the mission scientists find the craft has strayed, they can fire rocket thrusters to correct any drift. Pioneer 10 and 11, on the other hand, were going to keep themselves stable using the same trick that keeps a child's spinning top upright: they were going to spin their way through space. The spin provides a force that fixes the top's orientation; on Pioneer, the spin meant the mission scientists wouldn't have to worry about firing any thrusters to keep the craft on track.

Anderson realized that, since they were traveling under the influence of gravity alone, the Pioneer trajectories would provide a perfect test of gravity's nature. He submitted a proposal to NASA to use the probes for this purpose as well as their main mission, the investigation of Jupiter and the outer solar system. The NASA authorities agreed it would be a good test, and funded the extra experiments.

The first Pioneer probe was launched from Cape Canaveral on March 2, 1972. Pioneer 11 went up on April 5, 1973. Another seven years passed, years in which Richard Nixon resigned, Saigon fell, and Margaret Thatcher became prime minister of Britain. And then John Anderson noticed something odd.

Through all the years of their journey, the instruments on board the Pioneer probes had been sending back their readings to Earth. In 1980, the trajectory readings stopped making sense: both spacecraft, it seemed, were being pulled toward the Sun. Anderson talked to a few astronomers within his team about the anomaly, but he didn't go public because he couldn't explain it. Then, in 1994, he took a phone call from a physicist based at the Los Alamos National Laboratory in New Mexico.

Michael Martin Nieto was on a mission to find out just how reliable our gravity theories were. Whenever he came across other physicists, he would ask them what seemed like a dumb question: Can we still predict the motion of things using Newton's inverse square law if they lie outside our solar system? Eventually, he spoke to someone on Anderson's team, who said it might not be such a dumb question—and that he should ask John Anderson's opinion. Nieto made the call.

"Well, there is this Pioneer thing," Anderson said.

Once he had picked himself up off the floor, Nieto began to talk widely about the issue. Which is how Slava Turyshev got the Pioneer bug.

Turyshev has the distinction of being the first Soviet scientist to be employed at NASA's Jet Propulsion Laboratory (JPL) in Pasadena, California. When he came across Nieto's story, he had been invited over to do some work on his specialist subject, Einstein's general theory of relativity, the equations that describe how matter and energy shape the universe. He was only supposed to be in California for a year, and he thought that would be plenty of time to sort out this Pioneer nonsense. Fifteen years later, he is still there—and heading the investigation into the anomaly.

IF he had followed his first love, Slava Turyshev would have ended up an engineer, not a theorist specializing in general relativity. He grew up in a remote region of the Altai Mountains in what is now Kazakhstan; Turyshev's childhood was spent within viewing distance of the cosmodrome at Baikonur, the place where human spaceflight began. It was from here that Yuri Gagarin had been hurled into space in 1961. This was the 1970s, and the Soviets had become expert in spaceflight. From the balcony of his family home, the young Turyshev would watch in awe as the needle-sharp rockets pierced the sky. On treks up into the mountains, he and his father would sometimes come across shattered metal debris. He knew exactly what it was; he had watched the second-stage rockets being jettisoned in a cloud of gas a couple of minutes after launch, and falling back to Earth like Lucifer expelled from heaven.

Inspired by the Soviet space program, he and his friends began to make their own rockets. Turyshev, now in his forties, is proudest of "Ultraphoton," a two-stage rocket he built with his cousin. It was seven feet tall and was powered by a homemade gunpowder charge: sulphur scraped from scavenged matches. A glass Christmas tree bulb provided a suitable container for the charge; the ignition spark came courtesy of a 4.5-volt battery at the end of a one-hundred-foot length of wire. The launch was spectacular, he says. The heartbeat of his passenger—the young Turyshev's pet mouse—must have gone off the scale.

Everything was shaping up for Turyshev to become a space engineer. But when he was sixteen, someone showed him the equations for Einstein's general theory of relativity. And that was that. Somehow building rockets suddenly seemed a childish passion; the warp and weft of space and time, the mysterious fabric on which planets and people played out their dramas, seemed a far more fitting object for his attention.

By 1990 Turyshev had equipped himself with a PhD in astrophysics and theoretical gravity physics from Moscow State University. Three years later, he left for California.

TURYSHEV first came into the Pioneer project as the fixer—the cleaner. Like Harvey Keitel's character in *Pulp Fiction*, he was there to clear up the mess after people had done something stupid. Something stupid, in this context, was to have forgotten to factor in some subtle but important aspect of general relativity, Einstein's gravitational theory, in the planning of the Pioneer missions. But, to his surprise, Turyshev couldn't find anything wrong. And that is how his ongoing obsession with solving the Pioneer problem began.

Anderson, Nieto, and Turyshev all think they must have missed something. They don't want to rewrite the laws of physics; they want to leave Newton and Einstein alone. The trouble is, a massive analysis has failed to find anything on the spacecraft that could be causing it to drift off course. In 2002 they published a fifty-five-page paper together, going through everything they could think of to explain the drift. Nothing fit. And that was after Turyshev's cleaning job that checked every possible tiny effect of general relativity. Which came after Anderson's decade-long solo effort to find the problem. Something is pulling on the Pioneer probes with a tiny—but constant—pull. And, after nearly thirty years, it remains a mystery.

That is why, in several places around the world, researchers are watching the Pioneer probes fly all over again. It was Turyshev's idea to gather all the flight data from the Pioneer probes and write them into a computer program: Pioneer, the simulator.

It is a hugely demanding project. To understand why, think back to what information technology was like in 1973. Dot matrix printers are still new—and pretty cool. Bill Gates is still at Harvard; he is messing around with the university's computers but he hasn't yet dropped out to form a little company called Microsoft. That's still two years away. The first eight-inch floppy diskette drive had been invented just two years earlier. Which means that the Pioneer craft, designed in the 1960s, would store most of its data on the old-style punch cards. The mission data that aren't on punch cards are on rudimentary magnetic tape, coded in various programming languages that are the computer industry's version of ancient Latin.

Turyshev's problems don't stop there. NASA doesn't exactly archive all its mission data with loving care. These are records of when a thruster fired, or in what direction a spacecraft was pointing at 2:30 a.m. on a cold Friday morning in the early 1970s—they are hardly critical data. Unless, of course, the data challenge the laws of physics. But who knew that was going to happen?

No one at NASA, obviously. Turyshev eventually found most of the Pioneer trajectory data—four hundred reels of magnetic tape recording the computer's logs of the missions' paths through space—in a pile of cardboard boxes under a staircase at JPL. The tapes had suffered decades of neglect, heat, and humidity, but colleagues helped him restore the data and rerecord them onto DVD. Next, he went in search of the records from the onboard instruments that would reveal every move and spin of the Pioneer probes. He eventually found them at NASA Ames, in Moffett Field, California: sixty filing cabinets' worth of instrument readouts. They had been earmarked for imminent destruction.

The administrators at Moffett Field needed the space the filing cabinets were taking up, and were about to dump them in a landfill. Outside, in the parking lot, the first dumpster was waiting to be filled. In a moment of passion, Turyshev told them the discs were too important to throw away; he would rent a truck and take them away himself. The administrators were impressed and let the discs stay. They are now on DVD too. And all these data have been distributed to interested parties around the world. The refly of the Pioneer probes is going to be a global effort.

EVERYONE involved in the refly thinks the solution to the mystery will be something onboard the craft. After all, it wouldn't take much—just 70 watts of heat, for instance, could explain everything. As the heat radiation escapes, Newton's equal and opposite reaction would push the probe in the other direction.

The probes do indeed carry a source of heat: the probes' radioactive plutonium generators that power the crafts' electrical systems. When the probes were launched, these generators, stuck on long booms at the side of the craft so as to minimize any radiation damage, produced 2,500 watts of heat. Even now, they could produce 70 watts.

They could. But if they did, it would push the probes in the wrong direction. The generators are mounted at the side of each craft. To produce the anomalous acceleration toward the Sun, they would need to be mounted on the front.

There's a long litany of ideas like this—plausible mechanisms that have all been ruled out after careful examination. The software has all been checked, too; there are no faults that would result in a false reading of the trajectory or a slight shove off course. A fuel leak could do the trick, but it would have to be one that happened on both craft, in exactly the same way, and was not picked up by the internal instruments on either craft.

After three decades of trying to find an answer, the researchers investigating the Pioneer anomaly have nothing. If it's frustrating, it's also intriguing—so intriguing, in fact, that even NASA's head honcho, Michael Griffin, has become interested. Turyshev has had a number of conversations with Griffin about Pioneer. Maybe that's why, after years of studying Pioneer in their spare time, NASA researchers now have money for the project.

And rightly so. From the start, the Pioneer investigators have been almost exemplary when dealing with things that don't make sense. They won't embrace the extraordinary until they rule out the ordinary. Turyshev is almost pathologically opposed to talking about the exotic physics ideas, even the tamer ones, like a modified version of Newton's law. Nieto is the same. He is proud of all the Pioneer investigators have achieved so far, all the possible explanations they have ruled out. And his gut feeling is that the explanation for the Pioneer anomaly will turn out to be something like forgetting to turn off the lights. Or whatever is the NASA equivalent.

EVERY month, one or two new papers appear that espouse some exotic explanation for the Pioneer anomaly. The arguments often appear slightly unhinged; perhaps, for instance, the expansion of the universe caused the clocks involved in the measurements of the Pioneer probes' position to accelerate relative to each other? If that were true, Einstein's special relativity would require the analysis to be redone. The trouble is, this kind of outlandish phenomenon (and more than one has been offered) would also affect the motions of the outer planets, and these planets are not doing anything odd.

Or maybe the signal photons, the particles of radiation that carried information from the craft, had their wavelengths altered by the expansion of the universe? The researchers offering this suggestion admit that it fails a crucial test: it would push the apparent position of the Pioneer probes the wrong way. Perhaps the anomaly has to do with the signal photons having their quantum states shifted, or their being accelerated according to the laws of *nonlinear electrodynamics*, a theory developed in 2001 by a pair of Brazilian physicists? Or maybe the answer lies with John Moffat's extra universal force, the force that would also explain dark matter? The proponents of MOND think their theory also explains the Pioneer anomaly. Or, depending on which way you want to see things, is backed up by it.

Nieto disagrees. The MOND hypothesis doesn't tie in with the Pioneer data, he says; it doesn't produce the right kind of drift. He is OK—more OK than Turyshev, at least—with all the speculation. He wants to push the boundaries; he wants to know more than we know at present. But not at any cost; he understands the dangers of scientists wanting something extraordinary to be true. "If you go into it believing you're going to find something— oh God, you are in for trouble," he says.

In the end, Nieto believes they will find a straightforward explanation for the Pioneer anomaly. He is not deflated by this prospect, he says—not at all. We will have gained innumerable analysis techniques, and experience of handling data with exquisite precision, he points out. We will know the anatomy of a spacecraft—and of the space and time it travels in—with an intimacy that we never would have gained without Pioneer.

And if he's wrong—if all that effort reveals a force that is new to physics—so much the better. "For science it's a win-win," Nieto says. Anderson also thinks the Pioneer anomaly is most likely a false alarm. But he is leaving a door open for something revolutionary because he can't help but notice the parallels with another anomaly, one that Einstein inadvertently solved when he came up with general relativity.

IN 1845 Urbain Jean Joseph Leverrier, the French astronomer best known for the discovery of Neptune, calculated that Mercury's elliptical orbit around the Sun would experience a shift in its *perihelion*, the point of closest approach to the Sun, with each revolution.

This shift, or *precession*, is due to the gravitational pull of the other planets in the solar system. It is not unique to Mercury; the perihelion of every planet's orbit exhibits a similar precession. Mercury's, however, was not what it should have been. When Leverrier worked out, using Newton's laws, how big the shift should be, it didn't match the value astronomers had worked out from their observations. The discrepancy was forty-three seconds of an arc—just a little more than one hundredth of a degree—per century.

Noticing such a tiny anomaly was a hugely impressive feat for the time, equivalent to measuring the diameter of a penny from thirty miles away. But no one was patting themselves on the back; faced with the discrepancy, the scientists had no choice but to find an explanation. Astronomers tried various ad hoc fixes. Leverrier, perhaps inspired by the way he had been able to predict Neptune's existence by reference to other planetary orbits, thought the Mercurial discrepancy must be a sign that there was another planet waiting to be discovered. Others suggested the Sun had some kind of uneven weight distribution, or that dust clouds in between the Sun and Mercury were affecting the orbit. Nothing worked. It was only in 1915, when Einstein pointed out that a massive object like the Sun would warp the space around it, that an explanation for the anomaly was found.

Using his equations for general relativity, Einstein worked out that the warp in space, added to the tug of the other planets, would give a value for Mercury's perihelion precession of 42.9 arc seconds per century. It was a weighty validation for Einstein's newly minted theory and led to its immediate acceptance. And, according to John Anderson, it's a lesson for those who would discount the potential impact of the Pioneer anomaly.

If the explanation for the Pioneer anomaly is mundane, Turyshev's careful approach will almost certainly find it. If the explanation is something extraordinary, however, even the most meticulous sifting through the landscape of dull possibilities won't help. Mercury has taught us that ruling out the ordinary is not always going to lead to the answer.

Perhaps Pioneer doesn't offer enough data to build a picture of another force in the universe, Anderson says. But even if no one uses the errant flight path to create a breakthrough in physics, Pioneer could at least provide the validation for a theory developed by other means. Einstein didn't create general relativity because of the problem with Mercury's orbit, but the problem was hugely significant in proving Einstein's radical ideas were right. If the orbit of Mercury provided the perfect validation for one of the most important breakthroughs in science, perhaps the Pioneer spacecraft will one day do the same.

IS some unforeseen breakthrough coming? So far we have gathered evidence that the constituent parts of the universe are largely unknown, that the four-hundred-year-old law of gravitation could be in need of a rewrite, and that an unknown force might be responsible for pushing two of our spacecraft—craft that were predicted to offer a test for Newton's law of gravitation—off course. Kuhn might call this a sign of impending crisis. It certainly seems, as the foundations creak a little, that our current picture of the cosmos might have to change in the near future.

It's an exciting thought, but it doesn't allow us to say anything concrete about the future of science. All we can do is press on and add a new finding to the pile of evidence.

3

VARYING CONSTANTS

Destabilizing our view of the universe

Flap your arms and see if you fly. Chances are, you won't. The downward pressure of your arms on the air, and the equal and opposite reaction upward, are not enough to lift your body weight against gravity. The exact figures involved come from Newton's universal law of gravitation. (Whatever its accuracy over cosmological distances, it works just fine here.) The lift you would need to generate for takeoff involves the mass of the Earth, your mass, your distance from the center of the Earth, and a number known as *Big G*.

Newton's equation arose from the simple observation that two masses pull on each other, and Big G is a measure of how strong that pull is. The interesting thing is, there is no rationale for that number, no explanation for why Big G has the value it does. Scientists have worked out its value from experiments that balance the gravitational pull against another known force, such as the centrifugal force that wants to throw Earth out of its orbit, but just as scientists don't know where gravity comes from, they also don't know why it should have the strength that it does.

Big G has another, more scientific name: *the gravitational constant.* It is probably the most familiar of the fundamental constants of physics, the collection of numbers that describe just how strong the forces of nature are. Though every one of their values is derived from experiments, not from some fundamental understanding, they are integral to what we call the laws of physics: the constants make the laws work when we use them to describe the processes of nature. And because we assume that flying by flapping our arms will be as difficult tomorrow as it is today—that is, we assume that the laws of physics are immutable, eternal—we have to assume the constants don't change either. Which is why John Webb has got himself into such trouble.

The laws and constants have helped us define and tame the natural world. But what if there are no immutable laws? What if the constants aren't constant? Or, as Webb puts it, a wry smile playing across his lips, "Who decided they were constant, anyway?"

WEBB is a professor of physics at the University of New South Wales in Sydney, Australia, but his first encounter with this question came while he was a graduate student in England. One of his professors, the cosmologist and mathematician John Barrow, suggested they resurrect a question first raised in the 1930s by the British physicist Paul Dirac: Have the laws of physics remained the same for all time?

What is known as the *standard model* of physics inserts something like twenty-six numbers in its equations in order to accurately describe the strengths of the various forces in nature. The values we have for those numbers come from experiments done on Earth, and mostly in the twentieth century. Who's to say whether the same experiments done on Alpha Centauri, or 10 billion years ago, would give the same result?

If you want to check whether something has been the same for a long time, you need a sample that's as old as possible. Webb and Barrow quickly realized they had access to a perfect sample: the light emitted, 12 billion years ago, by quasars, the hearts of young galaxies. The emission of light from a star involves a constant that is officially known as the *fine structure constant*, but is more often referred to as *alpha*. The quasar light would depend on alpha as it was 12 billion years ago, so analyzing that light would provide the best possible chance of answering Paul Dirac's question. By 1999 John Webb had what looked like an answer.

The photons of light that carried his answer had traveled 12 billion light-years across the cosmos and landed on Earth in Hawaii, at the Keck Observatory that sits on the summit of Mauna Kea. But what was most interesting about the light arriving at the Keck telescope was the light that *didn't* arrive. Just as Vesto Slipher had done at the Lowell Observatory eight decades earlier, Webb and his team spread the light out into a spectrum. There were gaps in Webb's spectrum: his rainbow had missing colors. That wasn't interesting in itself; on a 12-billion-year journey through space, you'd expect the light to encounter some matter—clouds of gas are the usual culprits—that absorbs light of particular wavelengths. This leaves breaks in certain parts of the spectrum, as if a decorator has left a few vertical white stripes in the middle of your orange bedroom wall.

The interesting part of Webb's discovery was that the breaks were in the wrong place. Every atom, whether it is in an interstellar gas cloud or on the sole of your foot, will only absorb photons of particular energies. The energies in question differ for each atom; it is something like the atomic version of a fingerprint. As a result, by looking at the spectrum of light—and what is missing from it—you can fairly easily work out what atoms the light encountered.

The fingerprints in Webb's spectrum corresponded to two atomic encounters. One involved absorption by magnesium atoms; the other, by iron. It was clear from Webb's spectrum that the quasar's light had passed through clouds of magnesium and iron on its trip to Earth. But there was a problem. Although it was unmistakable which of the well-known absorptions the gaps in the spectrum were meant to correspond to, they were slightly out of place, as if someone had nudged the spectrum. For some, the absorption lines were nudged slightly to the left. For others, they were shifted a little to the right.

Webb sat down and redid the calculation. All the shifted lines made sense if he made one little adjustment. All he had to do was allow that when the light was racing through the interstellar dust clouds, the fine-structure constant was very slightly different from what it is today.

It sounds like a straightforward conclusion, but it took some guts to go public with the suggestion. Webb has been attacked for this; people, as he politely puts it, have "questioned his sanity" in remarking that a constant of nature might change over time. Especially one as central to physics as alpha.

ALPHA determines what happens every time a photon hits some piece of matter. Look at the wall opposite you; whatever color you see, you see because of alpha. A photon of light hits an atom in the paint. The atom absorbs the photon's energy and uses that energy to send out a photon that hits your eye. The energy of that photon determines the wavelength of the light it produces—in essence, what color you see. If the wall is orange, the photon has one energy; if it is violet, the energy is very slightly higher (it is still only equivalent to the energy in a billionth of a billionth of a raisin). To work out what color you'll see from a particular paint, you need to do a calculation that invokes alpha and the quantum structure of the atoms and molecules in the paint.

On the face of it, alpha is just a number. It is, roughly, 0.0072974, or 1/137 if you prefer fractions. The recipe for this number is fairly straightforward (though it depends on what units you're working in). First, multiply the charge on an electron by itself. Then divide that by a number called *Planck's constant.* This is a staple of quantum physics; physicists refer to it simply as h, and it describes the relationship between a photon's energy and the wavelength—the color—of its light. Next, divide what you've got by the speed of light. Now multiply the whole thing by 2π. Now you have alpha.

The thing is, alpha is not just about interior decoration choices; it is a pillar of physics and central to our entire description of the universe, beginning to end. Alpha determines how much energy there is in "empty" space, dictating how the newborn universe would expand. Once the first three minutes were over, alpha came into play in the electromagnetic interactions between the newly formed protons: it determined what kinds of photons filled the void.

When the first stars formed, as hydrogen atoms collapsed together and their nuclei fused under the intense gravity, alpha determined how much light and heat they gave out. And since radiation of all kinds give us our only view of the early universe, alpha tells us almost everything we know about the story of the cosmos. It might be made of nothing more than the speed of light, a rather boring number from quantum theory, pi, and the charge on an electron, but it is tied in to almost every process in the universe. Which makes it all the more unsettling that it might once have had a value that's different from the one we currently assign it.

Alpha's significance is due to the fact that it is the most important constant in one of our most important theories of physics: *quantum electrodynamics*, or *QED*. This governs any and every interaction between the charged subatomic particles: the protons and electrons. QED brings together quantum theory, relativity, electricity, and magnetism to describe the origins of electromagnetism. Alpha is also linked, via the "electroweak theory" that gained Steven Weinberg, Abdus Salam, and Sheldon Glashow the 1979 Nobel Prize in Physics, to the "weak force" that gives rise to phenomena such as radioactive decay in atomic nuclei. Since electromagnetism and the weak force are two of the four fundamental forces of nature, it is fair to say that alpha plays a pivotal role in the universe.

Not that the theory provides a value for alpha; scientists have had to do intricate experiments with electrons to work out what number they should plug into the QED formulas. Just as experiments gave us the gravitational constant that tells us how much the Earth and the Sun pull on each other in Newton's theory, the experimentally sourced alpha tells us how strongly charged particles affect each other. And it is not allowed to change by much.

Tweak alpha too far, and small atomic nuclei—those of helium, for example—would blow apart as the protons repelled each other. Stars wouldn't shine. Grow alpha by 4 percent, and the stars wouldn't have ever produced carbon—and thus we wouldn't exist.

Not that John Webb wants to change alpha by quite that much. Webb's absorption lines all makes sense if you allow it to have been smaller by just a millionth of its present value 12 billion years ago.

It seems, on the face of it, an almost inconsequential correction. A constant of physics, one that hardly anyone outside the subject has heard of, may have had a slightly different value in the past. It's put on a little weight, got one-millionth bigger in 12 billion years. Big deal. But it is a big deal. If it is true—and ten years later Webb still prefaces all his statements with this cautionary clause—if it is true, it opens a door to all kinds of unsettling ideas. We have built our story of the universe, and our explanations of how everything behaves within it, on the premise that the constants are, and always have been, constant. And, as we have seen, if the constants change, so do the laws. John Webb's observations are threatening to unleash a lawless universe.

Webb knows this; he is not rushing in to make any claims. He is an astonishingly careful man. He has already spent nigh on a decade trying to find the fault with his own results. His research team have dissected every result, carried out ruthless and rigorous statistical analyses, checked everything for some casual error. They have found nothing wrong. In fact, their analyses have taken them to the point where the varying alpha result has much more credibility than is generally required in any other area of physics. You don't even need Webb's level of certainty to claim a Nobel Prize for the discovery of an entirely new particle.

Nonetheless, most of the discussion about Webb's results tends to be about how they must be wrong—how there must be some error in the analyses. So, can we check? The obvious thing to do is to look at Webb's claim about alpha using something other than starlight and telescopes. The trouble is, you can't redo Webb's work in a simple laboratory experiment because it has to do with alpha's variation over a cosmological timescale. You can't measure how light interacts with matter in June, July, and August, find a consistent result every time, and claim Webb is wrong. He isn't claiming alpha is varying now; all he's saying is that it was very slightly different 12 billion years ago. If you want to do an experiment to test Webb's suggestion that alpha was different in the past, you need some evidence from the distant past. Fortunately, though, there is a way to get some: take off your lab coat, put on a pith helmet, and head into colonial Africa.

GO to eBay's French site, and type in the word *Brazza*. Chances are, the word means very little to you, but you'll bring up a range of collector's items for auction: matchboxes, pens, portraits, and cigars, to name but a few. In 1880s Paris, Brazza merchandise was all the rage. Pierre Savorgnan de Brazza, the French explorer (he was Italian by birth, but the Italian navy couldn't satisfy his thirst for adventure), put the West African territory of Gabon into French hands. And that made him a French national treasure.

Although the French named the Congo's capital city after him, Brazza's status as a treasure didn't last his whole life. He had established the Gabon colony with extraordinary integrity—there was fair trade, no slavery, and no subjugation by force under Brazza's governorship. With Gabon's rich resources, it was a strategy that was bound to win him enemies, and he spent the latter years of his life trying to beat down the flames of corruption and slavery that had begun to spread through the colony. For his trouble, Brazza was smeared, vilified, and, according to his wife, eventually poisoned.

One of Brazza's last acts was to establish the city of Franceville in the far east of Gabon as a place to resettle former slaves. And it was near here, at Oklo, that French nuclear scientists made the extraordinary discovery that has had enormous repercussions for John Webb's work.

In 1972 Francis Perrin of the French atomic energy commission was examining samples of ore from a uranium mine in Oklo. At the time, France was constructing a host of new electricity-generating nuclear reactors to be powered by Gabon's bountiful uranium resources. The next task on the to-do list was deciding what to do with all the nuclear waste they would produce. That meant cataloging the waste to decide how radioactive it was and how it needed to be managed. During this work, Perrin couldn't help but notice that the Oklo ore samples looked exactly like nuclear waste.

Atoms of uranium come in several different weights, or isotopes. Perrin noticed that the Oklo samples contained twice as much of one isotope, uranium-235, as would normally be expected. It took a few calculations, some careful analysis of the region's geology, and a great deal of lateral thinking, but eventually Perrin declared—to almost universal derision—that Oklo had once been the site of a natural nuclear reactor. Two billion years ago, a combination of heat and groundwater movement had provided the perfect conditions for fission reactions to take place underground.

At the time, the French nuclear authorities thought some kind of contamination was more likely. Since then, however, more natural reactors have been found in the Oklo region, and Perrin's finding is now universally accepted.

To science, the discovery is a goldmine. Two billion years ago, the constant we call alpha was presiding over the precise mechanics of the nuclear reactions that took place in the ground at Oklo. If you want to know whether alpha really is constant, Oklo provides the best test samples this side of Alpha Centauri.

The physicist Freeman Dyson was one of the first to jump on Perrin's find. Dyson, who has the reputation of being something of a rebel, had already been wondering, like Dirac, whether constants and laws were really so unchanging. The Oklo reactor gave him a chance to find out. He enlisted the help of the French nuclear physicist Thibault Damour and set about the analysis. Their conclusion was probably disappointing to Dyson: if alpha had changed at all, it was by no more than a billionth of its present value.

When Webb's results came out, Dyson and Damour's Oklo data allowed most scientists to ignore him; Oklo contradicted Webb's findings and was much more reliable than an investigation of ancient starlight. Eventually, though, as Webb's findings refused to go away, a few people did start to look more closely at what Dyson and Damour had done—and they began to find flaws. There was no firm rebuttal of the Oklo evidence until 2004. But when it came, it was more than a rebuttal. It came down firmly in support of a varying alpha.

Steve Lamoreaux and Justin Torgerson of the Los Alamos National Laboratory in New Mexico, the site of the United States' Manhattan Project, used what Lamoreaux calls "more realistic" estimates of the energies involved in the various nuclear processes that would have occurred. And that's not just Lamoreaux's take; Damour has concurred that these calculations should take us closer to the truth. The conclusion? Alpha has decreased by more than forty-five parts in a billion since the Oklo reactor burned itself out.

The fact that alpha has decreased since Oklo, while increasing since the starlight passed through gas clouds 12 billion years ago, might seem contradictory. But, as the evidence for varying constants builds up, it seems that this disparity may be, in fact, part of a cosmic conspiracy.

IN 1935 the British astronomer Arthur Eddington published a manuscript titled *New Pathways in Science*. In it, he described what he called the four "ultimate constants" of nature. One was a number he had worked out during a transatlantic boat crossing: the number of protons in the universe. Another was alpha—or rather its inverse: 1 divided by alpha. The third was the ratio of the gravitational and electromagnetic forces that pull an electron toward a proton. The fourth was even simpler: the ratio of the proton's mass to the electron's mass.

The fact that he could use these four numbers—and these four alone—to describe the characteristics of the entire universe impressed Eddington; physics must be doing pretty well, he thought. But, being a physicist and a close friend of Albert Einstein, who was trying to produce a single "unifying" theory of physics at the time, Eddington was also frustrated by the fact that there wasn't just one number. "Our present recognition of four constants instead of one merely indicates the amount of unification theory which still remains to be accomplished," he wrote. It would probably bug him more to know, as we do now, that at least two of those "constants" appear to be inconstant.

The second inconstant constant revealed itself in light captured by the telescopes at the European Southern Observatory in Chile. In 2006 a team of physicists published a paper declaring that the ratio of the proton mass to the electron mass, usually referred to as *mu*, was bigger in the distant past. This time, the shift was registered by looking at how the light changed as it passed through clouds of hydrogen gas. Hydrogen is composed of a proton and an electron, and the way it absorbs and reemits the light gave the researchers a value for mu. The wrong value.

As with alpha, this is a very distant past and a very small change: mu was bigger by 0.002 percent about 12 billion years ago. It was a significant enough result to be published in the prestigious journal *Physical Review Letters*, however.

It is significant because the electron and the proton mass are central to determining the strength of the "strong" force that holds atomic nuclei together. The strong force also binds quarks, the constituents of protons and neutrons. Since alpha is linked to the "weak" force that governs radioactive decay and the electromagnetic force that dictates the power of electrical and magnetic interactions, that's three out of the four fundamental forces of physics (the other is gravity) that seem slightly wobbly.

How do we deal with this? Perhaps Webb has been living in Australia too long, but he has a simple answer: don't sweat it. While many—if not most—physicists don't react to observational evidence of varying constants because it is simply too frightening, Webb has a very different, though no less pragmatic, stance. Alpha was only declared a constant in 1938, he points out. Mu was declared to be constant in 1953. It's not even as if we know anything about *why* these constants have the values they do—and that includes the gravitational constant. No one can explain them; there is no deep theory that matches the constants to their experimentally determined values. And so there really doesn't seem to be a good reason to fiercely cling to the notion they *must* be constant. In 2003, in the magazine *Physics World*, Webb put the case for coolness like this.

> When we refer to the laws of nature, what we are really talking about is a particular set of ideas that are striking in their simplicity, that appear to be universal and have been verified by experiment. It is thus human beings who declare that a scientific theory is a law of nature and human beings are quite often wrong.

So, if we're not to panic, what conclusions do we draw? Webb and Barrow have thought long and hard about this. Maybe, they suggest, the varying constants are telling us something. The fact that alpha seems to vary in different ways—smaller than now 12 billion years ago, but bigger than now a couple of billion years ago—suggests that the constants (and maybe the laws) could vary in both time and space. Perhaps, were we to wander through the vastness of the universe, we would come across different sets of constants and different sets of laws—parochial cosmic by-laws—wherever we went. It is a short step from there to suggesting that the laws are not fixed in time either. Maybe the laws of physics changed as the universe evolved?

This is not an entirely new idea. John Webb has been labeled incompetent or (more often) studiously ignored by his detractors, but all he has really done is uncover an anomaly that backs up the suggestions of one of the world's most respected physicists. Thirty years ago, the brilliant physicist John Wheeler asked why we assume the laws are unchanging. The strength of the forces of nature might depend on cosmic conditions, he suggested, making them different in the hot, dense plasma of the birth of the universe than they are in today's old, cold cosmos. Might the laws not change their character as the universe cools down, flowing then congealing like a metaphysical molten lava? It is a very loosely formed idea—Wheeler in fact called it "an idea for an idea"—but it raises the possibility that our attempts to trace cosmic history, from the big bang through the production of the first elements and stars, might be hugely oversimplified.

Richard Feynman, too, had his doubts about our grasp of the laws of physics. In 1985, twenty years after he, Julian Schwinger, and Shin'ichiro Tomonaga won a Nobel Prize for the development of QED, Feynman published a slim book on the theory. In the final chapter, titled "Loose Ends," he makes an honest admission that seems somewhat surprising given the theory's success and acceptance. "We do not have a good mathematical way to describe the theory of quantum electrodynamics," he says.

To give the quote some context, Feynman is pointing out that the coupling between light and matter depends on inserting a couple of numbers that are found through "hocus-pocus" rather than experiment. What's more, he says, you then also have to insert what he calls "one of the greatest damn mysteries of physics, a magic number that comes to us with no understanding by man." He is talking about alpha, of course. Despite being one of the most successful theories of physics in existence, QED still has Feynman cursing—and mostly because of alpha. "It has been a mystery ever since it was discovered more than fifty years ago, and all good theoretical physicists put this number up on their wall and worry about it."

By the time Schwinger died, he had more reason than most to worry about alpha: an investigation into QED, the theory that invokes alpha, had all but scuttled his career. The investigation in question was carried out by two chemists: Stanley Pons and Martin Fleischmann. They are now almost universally derided as frauds, cranks, or—at best—incompetents, and Schwinger's resolute support for their work destroyed his hard-won credibility. For more than a decade, the fate of Pons, Fleischmann, and Schwinger has stood as a warning to others. Whatever the benefits and the insights it might bring—and they are, potentially, legion—scientists investigate our next anomaly, known as cold fusion, at their own risk.

4

COLD FUSION

Nuclear energy without the drama

*SALT LAKE CITY—Two scientists have successfully cre-
ated a sustained nuclear fusion reaction at room temperature in a chemistry laboratory at the University of
Utah. The breakthrough means the world may someday
rely on fusion for a clean, virtually inexhaustible source
of energy.*

That was how a press release, issued on March 23, 1989, by the University of Utah, launched the end of Martin Fleischmann's career. Fleischmann remembers his work's motivation very differently. "I had no intention of saving the world," he says. "No intention whatsoever!"

Fleischmann speaks with a vaguely eastern European accent—he was born in Czechoslovakia—but he doesn't speak a great deal. Ask him a question, and he is quite capable of sitting, musing on it, for a full minute or more. Perhaps he has learned caution since that day.

He has a lot of regrets about that press release, and the press conference that followed it, but the one that he admits first is that he never told the truth. "I never told people I was only interested in understanding quantum electrodynamics," he says.

It was the summer of 2007 when I met Fleischmann for the first time. Just coming face-to-face with this man, now a curiosity in the history of science, was a coup. His partner in the Utah experiment, Stanley Pons, lives in the south of France and sees no one—especially not journalists. Fleischmann, now in his eighties, is still fairly guarded about the outside world, and my visit only came about through a network of contacts. I am in good company, though. In the months after the March 1989 announcement, the Nobel laureate Julian Schwinger also tried and failed to set up a meeting with Pons and Fleischmann. In exasperation, he even sent a plea for a rendezvous in a letter to the *Los Angeles Times*. Eventually, a friend managed to set things up, and Schwinger got to go to Salt Lake City, where the three physicists sat and talked at length about the limits of the theory that had won Schwinger his Nobel Prize.

Fleischmann was also a visitor to Salt Lake City; Stanley Pons was the Utah resident, and it was in his lab that the room-temperature fusion—now infamous as *cold fusion*—experiments had taken place. Together, Fleischmann and Pons had plowed about $100,000 of their own money into the experiments, but they had hit a brick wall. They needed another $600,000 to progress. They wrote a grant application, in which they mentioned how an improved understanding of the processes of nuclear physics—in particular, how nuclear energy might be released in room-temperature reactions— might allow you to create a new source of power. Put simply, you could get more power out than you put in, as with an atom bomb, but with a lot less drama. It was this that the university seized upon when it strong-armed Pons and Fleischmann into announcing their results at a press conference: the university's research was going to save the planet. Fleischmann was mortified, but—to his enduring regret—played along. His complicity cost him his reputation and his career. For a couple of weeks, the world went mad for the story. Then the whole thing disappeared in a puff of scandal, partly because no one could replicate their results, but mostly because the results they claimed made no sense.

Nuclear fusion is real enough. Squash two atoms close enough together, and their nuclei join, or fuse, creating one heavy atom and releasing energy.

This is the source of life on Earth: the Sun is powered by fusion. In the Sun, hydrogen atoms are squashed together by the enormous gravitational pressure to make a single atom of helium. This releases fistfuls of energy; small wonder, then, that scientists have long dreamed of creating controlled nuclear fusion on Earth.

To make sunshine on Earth, the idea is usually to squash together "heavy" hydrogen atoms. Normally, hydrogen has no neutrons in its nucleus, but some hydrogen atoms contain one neutron (deuterium) or even two (tritium), making them heavier. These heavy hydrogen atoms are better for fusion than normal hydrogen because they will fuse at a lower temperature and pressure. In the Sun, fusion reactions take place at 10 to 15 million degrees and at pressures one hundred times the pressure in the deepest part of the ocean. On Earth, both the temperature and pressure conditions—which are necessary to overcome the electrical repulsion of the positively charged nuclei—are enormously hard to achieve. Any help, by using heavy hydrogen, for instance, is most welcome.

It's especially welcome since deuterium and tritium are easily available in seawater. In theory, there's enough energy in the oceans to meet all our needs. The reality is not quite so straightforward, however; researchers have been trying to perform controlled fusion reactions for a few decades. It's almost a running joke, in fact: whenever you ask about progress, the project is always a few decades from success. It's not clear we'll ever be able to create the temperature and pressure conditions of the Sun on Earth.

And that's what makes Pons and Fleischmann's claims so extraordinary. They implied that all the decades of effort and the millions of dollars of research money were perhaps a waste of time, that you could create fusion reactions and release nuclear energy at room temperature and normal atmospheric pressure—and in nothing more complex than a laboratory beaker.

Pons and Fleischmann's equipment was simple, to say the least. Their beaker contained heavy water, where each oxygen molecule was bound to two deuterium atoms rather than two simple hydrogen atoms. Into this they put one end of a rod made of palladium metal. The other end of the rod was hooked up to one side of a battery. The battery's other terminal was linked to a coil of platinum wire that spiraled around the inner wall of the beaker.

The setup meant that current from the battery traveled along the platinum wire, through the heavy water, and into the palladium rod. Pons and Fleischmann claimed that this resulted in deuterium atoms being packed into the spaces between the palladium atoms in the rod—and packed so tightly that they began to fuse, liberating energy.

The first part of the explanation makes some sense, at least. The Scottish chemist Thomas Graham was the first to note, in 1866, that palladium was able to absorb hydrogen gas. In fact, it seems to have an unusual appetite for the stuff. At normal temperatures and pressures, palladium can absorb nine hundred times its own volume of hydrogen. But would a palladium rod really absorb so much hydrogen that the atoms would begin to fuse?

Pons and Fleischmann claimed this because, they said, the experiment produced an anomalous amount of heat. The temperature of the water in the beaker rose far above anything explicable by the power coming from the battery. Energy was coming from somewhere, and the only possibility was the fusion of deuterium atoms.

When the pair first made these claims, there was a frantic race to replicate their results. The U.S. Department of Energy convened a panel of top-flight scientists—the Energy Research Advisory Board (ERAB)—to judge the outcome. In November of 1989, the panel brought its verdict. "Some laboratories support the Utah claims of excess heat production, usually for intermittent periods, but most report negative results," the report said. The panel concluded that the experimental results on excess heat "do not present convincing evidence that useful sources of energy will result from the phenomena attributed to cold fusion . . . the present evidence for the discovery of a new nuclear process termed cold fusion is not persuasive." As a result, the panel "recommended against the establishment of special programs or research centers to develop cold fusion." The most positive thing the panel had to say was that "some observations attributed to cold fusion are not yet invalidated." As a result, it was "sympathetic toward modest support for carefully focused and cooperative experiments within the present funding system." With most scientists baying for Pons's and Fleischmann's blood, it was never going to happen; people weren't even going to risk asking for money. As the writer Bennett Daviss put it, cold fusion was "as respectable in science as pornography in church."

There was one place where cold fusion wasn't thought of quite so poorly, though: the laboratories of the U.S. Navy's Office of Naval Research. Martin Fleischmann was a consultant for the navy, and plenty of the navy's researchers had published papers with him and were investigating their own lines of attack on the idea of low-temperature nuclear reactions. They knew very well that Fleischmann was no flake. Three years earlier, he had been elected a fellow of the Royal Society, the British academy of science that honors the most eminent scientific minds in Britain and the Commonwealth. He had published hundreds of peer-reviewed papers and had a reputation as one of the world's best electrochemists. When the Pons and Fleischmann furor broke, U.S. Navy researchers were asked by their superiors if anyone was working on something like it. Dozens of people put their hands up. And they were allowed to keep going.

It was kept on the down-low; the words *cold fusion* were nowhere to be found on the navy's budget sheets. The money came out of "miscellaneous" expenses and was marked up as supporting research into "anomalous effects in deuterated systems." Nonetheless, there was room for the navy's chemists to carry out their own investigations. Look back to the November 1989 report by the U.S. Energy Research Advisory Board, for example, and you'll find a contribution from Melvin Miles.

Miles's story is almost a microcosm of the cold fusion story. He is now retired from the navy, but in 1989 he was working in the laboratories of the Naval Air Warfare Center at China Lake in California. The author of a hundred or so peer-reviewed papers, Miles was no stranger to experimental rigor and figured he could test the cold fusion claims as well as anyone else. It was a decision that eventually brought his career to a humiliating end.

Miles's paper that was cited in the ERAB report is his take on a very straightforward piece of experimental science. Miles found a piece of palladium in his lab, which he dutifully soaked in deuterium for a week. The idea was that the palladium would become "loaded" with deuterium. Then he put the sliver of metal into an electrochemical cell and turned on the juice. Nothing happened. No strange heating effects, no evidence of nuclear reactions. Nothing. That's what Miles reported, adding his own research findings to the growing pile of evidence against Pons and Fleischmann.

He probably would have walked away then, but some of his colleagues—

people he respected—were still reporting occasional flashes of excess heat in their experiments. So Miles tried again. From March through August 1989, there was no change in the outcome. Then Fleischmann sent him a recommendation. Fleischmann's palladium samples were "Johnson Matthey Material A." Miles sent for some and tried them out. He published the results in the *Journal of Electro-Analytical Chemistry* in December 1990. In eight experiments, the new palladium samples produced somewhere between 30 and 50 percent more energy than he put in.

The paper conveys no sense of the excitement it ought to have generated. No media picked it up, but Miles was reporting, essentially, that he had repeated Pons and Fleischmann's experiments, gaining similar results. Not that his cautious reporting was enough to save his career.

Until 1996, Miles was relatively safe. His boss at the Office of Naval Research, Robert Nowak, was a chemist who allocated the cold fusion program a modest budget and defended it in the face of threats and mutterings from skeptics who didn't like federal funding to fall into the hands of cold fusioneers. Nowak also defended it in the face of failure: from 1992 to 1994 Miles never managed to repeat his generation of excess heat. The navy metallurgist supplying his palladium alloy electrodes used up the next two years—and all the manager's goodwill and patience—getting the recipe right. By the time he did get it right, producing electrodes that gave Miles a consistent 30 to 40 percent extra energy gain, the cold fusion budget had been canceled.

Most of the cold fusioneers managed to get work on other projects, but not Melvin Miles. Nowak left for a job at the Defense Advanced Research Projects Agency, and his successor told Miles that he was effectively unemployable. Everything had to be paid for, including Miles's time, and in the new climate no one wanted to buy the time of a researcher who had sullied himself with cold fusion research. The hundreds of peer-reviewed papers with Miles's name on them meant nothing. He was reassigned to a job as a clerk in the stockroom. Thanks to his cold fusion research, Miles ended his naval research career taking boxes down from shelves. The lesson? Involvement with cold fusion is a surefire way to blot your scientific copybook. It even happened to a Nobel laureate.

JULIAN Schwinger died in July 1994 from pancreatic cancer. Though it doesn't mention his cold fusion work explicitly, Schwinger's obituary in the journal *Nature* talks of a "bitter-sweet quality to the later part of his life." Noting that Schwinger refused to follow the newer directions and fashions in theoretical physics—they were "too speculative, inadequately linked to experiment"—he "became increasingly isolated and, to a degree, estranged from the world community of physicists."

Schwinger evidently saw the bitter more than the sweet. The reaction from his peers to his interest in cold fusion was mostly one of contempt. In 1991, three years before his death, he wrote, "The pressure for conformity is enormous. I have experienced it in editors' rejection of submitted papers, based on venomous criticism of anonymous referees. The replacement of impartial reviewing by censorship will be the death of science."

Schwinger's attitude toward cold fusion is summed up in a talk he wrote but never got to deliver; it was read to a conference on cold fusion five months after his death. "As Polonius might have said: neither a true-believer nor a disbeliever be," Schwinger wrote. "From the very beginning . . . I have asked myself—not whether Pons and Fleischmann are right—but whether a mechanism can be identified that will produce nuclear energy by manipulations at the atomic—the chemical—level."

Schwinger made several attempts to explain the cold fusion results and wrote eight theory papers. None of his theories properly explained the observations, but he never gave up; for him, it seems, the Pons and Fleischmann results provoked a fascinating question, one that he pursued until his death. Whether Pons and Fleischmann were right was not the issue; had they highlighted an issue worth investigating? Could nuclear energy be released by manipulating atoms in chemical processes? The man who had helped create a theory the *New York Times* hailed as "one of 20th-century physics' few unqualified triumphs" considered this a question worthy of his remaining years.

This fact alone makes it worth taking cold fusion seriously as an anomaly, and it is worth noting that some of Schwinger's early work, too, was driven by his interest in an anomaly. In the years shortly after the end of the Second World War, new experiments had shown that the "hyperfine" part of the atomic spectrum of hydrogen differed from that predicted by the standard theoretical model of the time, a model created by the British physicist Paul Dirac. Schwinger was enthralled—but cautious. The Harvard physicist Norman Ramsey, one of the experimentalists involved in highlighting the original anomaly, recalls that Schwinger didn't want to waste his time if it was all a fuss about nothing.

> Schwinger invited me to lunch and asked me searching questions about the reliability of the experimental hyperfine anomaly. He said he thought he could explain it, but would have to develop a relativistic QED; he was worried about doing all that work if the hyperfine anomaly wasn't real. I told him I was convinced it was real. He then worked vigorously on this problem.

On December 30, 1947, the journal *Physical Review* received an explanation for the anomaly. It required a novel combination of Einstein's relativity and the new theory of quantum electrodynamics. The journal duly published Schwinger's paper. It was the first application of *relativistic QED*, now an essential component of modern physics. If Schwinger was concerned to make sure the hydrogen spectrum anomaly was real before he invested too much time in the project, it seems likely he had also convinced himself that the cold fusion results were similarly worthy of his attention.

Science is not about people, though, and true anomalies stand by themselves because they don't go away. Cold fusion research has survived Schwinger's death, Miles's retirement, and Pons and Fleischmann's public excoriation: in 2004, a Department of Energy (DoE) study admitted that there might be something to cold fusion claims after all and recommended that "funding agencies should entertain individual, well-designed proposals" for cold fusion experiments.

This report was the result of the first examination of the evidence accumulated since the hastily compiled ERAB report of 1989. Things had certainly changed since then: the navy researchers, for instance, had released a two-volume report covering a decade of their research. What was perhaps most interesting, though, was how one of the original—and most damning—reports on Pons and Fleischmann's claims had been amended.

When Pons and Fleischmann first made their claims, there were three front-runners in the race to confirm or refute cold fusion. The results from MIT, Caltech, and the United Kingdom's Harwell Laboratory would be influential enough to outweigh results—positive or negative—from any other laboratories anywhere in the world. When all three of these heavyweights reported that they had failed to see any excess heat, it was all over for cold fusion.

The report from MIT wasn't exactly accurate, however. The MIT researchers have since admitted that their attempt to replicate Pons and Fleischmann's work did produce more heat than they had expected. Although the evidence never made it into the published report, an appendix added after publication documented excess heat.

This turnaround came to light after MIT's chief science writer, Eugene Mallove, received the MIT final paper. It was dated July 13, 1989, and showed no excess heat, a result that damned cold fusion. Mallove was then given an earlier draft of the same paper, detailing the outcome of the same experiments. It was dated July 10, and its data showed excess heat. In those three days, the data had apparently been changed from showing excess heat to showing none. Mallove lodged an official complaint, then resigned in protest.

Mallove's charges resulted in the appendix being added to the MIT report. It made no difference to the ERAB report because the report had already been presented to Congress as evidence that Pons and Fleischmann had no basis for their claims, but at least the record now shows that the heat graph had been altered. It had happened because the research team had decided that excess heat was not the smoking gun; it was a sudden release of heat that mattered, and their heat release had not been sudden enough. But it seems that they never had much confidence in their data either way; in Mallove's report about the affair, *10 Years That Shook Physics*, he recalls how Professor Ronald R. Parker of MIT's Plasma Fusion Laboratory publicly stated the calorimetry data were "worthless."

Calorimetry—the science of measuring heat—is known to be the hardest of sciences, and it is worth pointing out that calorimetry data are just as unhelpful today: according to the navy researchers, there is still no cold fusion experiment that has reliably and repeatedly produced a measurable excess of heat. Nonetheless, the last fifteen years of research have changed the picture enough for the DoE panel to concede there is something worth looking at in cold fusion. In the years since the DoE report came out, there has been a further breakthrough, too. The cold fusioneers now have reliable evidence that, whatever the calorimetry considerations, some kind of nuclear reactions are definitely going on in their experiments.

TO get energy out of atoms, you either have to break up their cores—a process called *nuclear fission*—or join different atoms together by nuclear fusion. Both processes liberate energy, but they also create a range of by-products that depend on what atoms you're using, and whether they are fusing or splitting. Many of those by-products are high-energy particles that shoot out of the reaction, and they can be detected.

Nuclear scientists use a plastic called CR39 to expose nuclear events. CR39 is the same kind of plastic that is used in eyeglass lenses. Place a piece next to a chamber containing nuclear reactions, and the particles flying out will break up molecular bonds in the polymer, creating a telltale pattern of microscopic pits and scratches. This pattern is like a fingerprint: if you know what you're doing, it's a fairly straightforward piece of detective work to look at the pattern and deduce what kinds of particles hit the plastic chip, and what energy they were traveling with. And that can tell you what kind of reaction was going on in the chamber.

Navy researchers have put CR39 chips—they look like microscope slides—into their cold fusion cells and given them to a couple of independent nuclear track specialists to look at. The specialists were convinced they were looking at the signature of a nuclear event. Put a CR39 chip next to a piece of depleted uranium, a radioactive metal, and it will become covered with random lines and concentric circles. Put one in a cold fusion experiment, and it ends up looking the same.

It may not sound like much, but the CR39 chips provide almost incontestable evidence that whatever is going on inside those simple experiments, nuclear reactions are involved. That's a big deal, and, as well as allowing them to come out and talk confidently to the navy's top brass about what they've been doing, the CR39 chip data have netted the cold fusion researchers their first publication in a major mainstream journal in many years. In June 2007 the findings were published in *Naturwissenschaften*, a journal that also published work by a certain Albert Einstein. The CR39 data have also convinced the navy to fund further research into cold fusion.

What they still don't have, though, is reliable evidence of extra energy. They make no claims of anomalous heat production or of nuclear fusion. In fact, they don't even use the *f*-word but refer to what is going on in their experiments as *low energy nuclear reactions*. In many ways, that is exceedingly frustrating: in cold fusion, calorimetry is everything—excess heat is what it is all about. Nonetheless, we have to accept what we have. For now, all the cold fusion anomaly has is the CR39 data. Maybe it will lead to a clean, virtually inexhaustible form of energy; maybe it won't. But we can say this: load palladium with heavy hydrogen molecules, pass a current through it, and some kind of nuclear reaction appears to take place.

One of the few publications to get immediate perspective on the original cold fusion debacle was the *Economist*. A month after Pons and Fleischmann's 1989 announcement, it said the affair was "exactly what science should be about." Even if the pair were wrong, there was no harm done; complaints about wasting time and money were cowardly reactions. Pons and Fleischmann had provided "excitement and inspiration in abundance." It seems almost laughably naive in the light of what followed, but the *Economist* was right: the research is what science should be about, and it has led us somewhere. What is clear, what is now more than a maybe, is that nuclear processes can be unlocked without a great drama of fire and storm. As we develop our understanding of nuclear physics beyond what's currently described by the theory known as quantum electrodynamics, the setup of the cold fusion experiments may one day prove to have been a fortuitous leap in the dark that catapulted us into a new age of nuclear science.

Perhaps Joseph Priestley has the most appropriate perspective here. During his lifetime, Priestley discovered oxygen and, by accident, invented carbonated water. "In this business," he once said, "more is owed to what we call chance—that is, to the observation of events arising from unknown causes—than to any preconceived theory." The story of cold fusion has been a debacle; it began with an attempt to probe a profound theory and has generated little more than scandal, exposing the worst sides of human nature (and the human nature of science). But it is not over yet, and there are signs that it could still yield something worthwhile, something that will eclipse its checkered history and make us glad that, before they became scientific curiosities, Martin Fleischmann and Stanley Pons were simply curious.

5

LIFE

Are you more than just a bag
of chemicals?

So far, we have looked at anomalies that have ranged from the grand scale to the smallest: from the ultimate nature of the universe to the nature of atomic nuclei. The implications have ranged from discerning the ultimate fate of the cosmos to harnessing a new form of energy on Earth. None, however, can be as fundamentally significant to humans as the implications of our next anomaly. It is so important that the Santa Fe complexity theorist Stuart Kauffman has said coming to grips with it could open the door to a whole new science. What is it? You know it best as the thing we call life.

In some ways, it's difficult to see life as an anomaly. But perhaps that is a contempt born of familiarity. Stop taking it for granted, and think for a moment about what sets the biological world apart from the world of nonliving matter. As scientific observations go, it's a cast-iron case: plenty of stuff has that quality we call *alive*. We also see plenty of stuff around us that no one would call alive. But no scientist on Earth can tell you where the fundamental difference between these two states lies. Neither can any of them take something from the not-alive state and turn it into something that everyone would agree is alive. In fact, scientists are still struggling to agree on what would even constitute such a step.

We are composed of molecules whose individual behavior and properties can be described by science—quantum theory provides the root explanation. Somehow, though, these molecules are put together in a way that results in properties that defy explanation by any theory. We recognize those properties as the thing we call *life*. But in many ways that's no more enlightening than the label *dark energy* is to cosmologists. As Erwin Schrödinger, the father of quantum theory, asked in 1944, "What is life?"

The answer that most scientists favor is "nothing special." There is no reason to believe something ethereal or spiritual, some "vital spark" switches on life in an assembly of molecules. There is also no reason to think the question is somehow beyond the scope of science, a mystical or philosophical phenomenon. There is, they say, no reason to think we can't find the answer. It's just that, at the moment, we're not sure where or even how to look.

There are many ways to try to unravel the essential nature of life. One is to find out how it started: trace the tree of life back to the point where all that existed was chemistry. Another is to try to build something that is "alive" from scratch: take some chemicals and put them together in ways that might make them come alive. A third option is to sit and think about what exactly it is that marks the difference between living and nonliving matter and come up with the definition of *life*. It is this latter path that is perhaps the most well trodden. It is also the one widely admitted to be a dead end.

How would you define *life*? Is it when a system reproduces itself? If that is the case, plenty of computer programs could be called alive, while plenty of people—sterile men and women, for example, or nuns—could not. Things that are alive consume fuel, move around, and excrete waste products, but so do automobiles, and no one would call them alive.

Schrödinger came to the conclusion that life is the one system that turns the natural progression of entropy, moving from order to disorder, on its head; living things are, effectively, machines that create order from disorder in their environment. This, to him, was the essence of the process that staves off the state of death. It is still not enough, though; a candle flame creates order from disorder in its environment and is patently not alive.

The physicist Paul Davies has perhaps done most to try to elucidate a definition of *life*, but he too remains stumped for a final answer. Instead, he considers life to have various characteristics, none of which defines *life* in and of itself, and many of which can also be seen in nonliving matter. In his award-wining book *The Fifth Miracle*, Davies lists these attributes—and their failings—as explanations or descriptions of life, rather than definitions. A living being metabolizes, processing chemicals to gain itself energy (as does Jupiter's Great Red Spot). It reproduces itself (but mules don't, and bush fires and crystals do). It has organized complexity—that is, it is composed of interdependent complex systems such as arteries and legs (in this way it is rather like modern cars). It grows and develops (as does rust). It contains information—and passes that information on (like computer viruses). Life also shows a combination of permanence and change— evolution through mutation and selection. Finally, and perhaps most convincingly for Davies, living beings are autonomous; they determine their own actions.

Others have added to this list. A living system must also be contained within a boundary that is part of the system, according to the biologist Lynn Margulis. Whichever way you look at it, though, the definition—or rather the series of suggestions and characteristics—is too vague to be really useful. In fact, attempts to define *life* are beginning to be seen as damaging. In June 2007 an editorial in the journal *Nature* declared that

> one might have hoped that such perceptions of a need for a qualitative difference between inert and living matter—such vitalism—would have been interred alongside the pre-darwinian belief that organisms are generated spontaneously from decaying matter. Scientists who regard themselves as well beyond such beliefs nevertheless bolster them when they attempt to draw up criteria for what constitutes "life."

The editorial was heralding the achievements of *synthetic biology*: the attempt to build life from its chemical components. This, in the establishment view, is the way forward for dealing with the fact that life does not fit into any existing modes of understanding. The question of whether it can succeed, though, is still wide-open.

THE first researchers to make a significant move toward creating life were the University of Chicago chemists Stanley Miller and Harold C. Urey. In 1953 they sealed ammonia, methane, hydrogen, and water in a flask to mimic the Earth's primordial atmosphere. Then they put sparks of electricity through the mixture. The idea was that lightning storms may have sparked primordial Earth's chemicals into creating the first life.

The experiment was an extraordinary success. After a week of continuous electrical discharge, around 2 percent of the methane's carbon had turned into amino acids, the building blocks of proteins. It was a revelation.

The trouble is, the experiment was flawed. The gases Miller and Urey used are not the ones scientists now think were present in the primordial atmosphere. In fact, the fundamental chemical characteristics of the mixture may have been entirely wrong. What's more, the stuff of Earth life— proteins, lipids, carbohydrates, and nucleic acids—didn't show up. The New York University chemistry professor Robert Shapiro likened the experiment's production of amino acids to the accidental production of the phrase *to be* during a random attack on a typewriter's keys; it doesn't mean the rest of *Hamlet* is going to follow. "Any sober calculation of the odds reveals that the chances of producing a play or even a sonnet in this way are hopeless," he says, "even if every atom of material on Earth were a typewriter that had been turning out text without interruption for the past four and a half billion years."

So, it's hard to call the Miller-Urey experiment a true success. Nevertheless, it showed what might be possible. And in 1961 the Catalonian Juan Oro went even further. Oro put water, hydrogen cyanide, and ammonia together and produced large amounts of adenine. Not only is adenine one of the four building blocks of DNA; it is also a major component of adenosine triphosphate (ATP), the chemical that provides the fuel for biology to work. You don't run, grow, or even breathe without using up ATP.

The Nobel Prize–winning Flemish biologist Christian de Duve once said that "life is either a reproducible, almost commonplace manifestation of matter, given certain conditions, or a miracle. Too many steps are involved to allow for something in between." If it really is that simple to make amino acids and adenine, perhaps it is easy to get life started. There is a good reason to take this viewpoint seriously: the astonishing rapidity with which life got going on Earth.

In the center of the Pilbara region of northwestern Australia, the Sun beats down on red rocks that were formed by the planet's first creatures. They are extraordinary formations, resembling egg cartons and upside-down ice cream cones, and their arrangement and shape tell us they were laid down as sediment excreted by microbes 3.5 billion years ago. Which means their shape is not the only extraordinary thing.

Our solar system formed just 4.55 billion years ago. For millennia afterward, it was a hellish maelstrom of asteroids and comets; huge rocks hurtled through space and pounded the planets and moons. According to our best ideas of how things came to be as they are on our planet, a rock the size of Mars slammed into the primordial Earth. The impact turned the planet's surface to liquid and sent a blob of molten rock into orbit—a blob that eventually became our silvery Moon.

The surface of Earth would have taken tens of millions of years to cool from that cataclysmic impact, and further impacts would have slowed the cooling. Studying the Moon's craters, formed only once the surface had hardened, tells us that the asteroid and comet storm only began to abate about 3.8 billion years ago. Only then could life begin; it seems it took the Pilbara microbes only around 300 million years to come into existence.

The cosmologist and astronomer Carl Sagan took the rapidity of life's emergence as proof that it can't be that hard to make. "As soon as conditions were favorable, life began amazingly fast on our planet," he wrote in an essay for the Planetary Society's *Bioastronomy News* in 1995. "The origin of life must be a highly probable circumstance; as soon as conditions permit, up it pops!" Sagan, who died a year later of myelodysplasia, a bone marrow disorder linked to leukemia, was led to conclude that life is extremely likely to exist elsewhere in the universe.

Many of today's biologists draw what is perhaps a more self-centered conclusion: if life pops up so easily, we ought to be able to make it. Most scientists working in this field agree that the task facing them is achievable; it is a matter of when, not if, they will create artificial life. After all, if it happened once—when a bolt of lightning happened to hit the right bowl of primordial soup—surely the collective power of today's biotechnologists can make it happen again. Launching Life 2.0 surely can't be that difficult.

Such bullish attitudes don't take account of our ignorance, however. For more than a decade, scientists have been sure they are on the cusp of working out exactly how life arose from its chemical constituents. But it's not clear that we're any closer to that achievement today than we were ten years ago. If creating life is "simply" a matter of putting the right chemicals together under the right conditions, there's still no consensus about what "right" actually is—for the chemicals or the conditions.

AFTER the first atomic bomb was tested in the desert near Los Alamos, New Mexico, J. Robert Oppenheimer, the project's chief scientist, made only one audible comment: "It worked." Yet, in an extraordinary piece of newsreel footage filmed years later, he admitted that his mind had been filled with much deeper thoughts at the time. Barely containing his emotions, looking down—almost at the floor—and wiping a tear from his eye, he recalls the moment.

> We knew the world would not be the same. A few people laughed, a few people cried, most people were silent. I remembered the line from the Hindu scripture, the Bhagavad-Gita. Vishnu is trying to persuade the Prince that he should do his duty and to impress him takes on his multi-armed form and says, "Now, I am become Death, the destroyer of worlds." I suppose we all thought that one way or another.

If ever there were to be another world-changing moment as profound as that bomb test, it would surely be the first time humans bring inanimate matter to life. In the middle of the New Mexico desert, Steen Rasmussen is attempting to do just that in his labs at the Los Alamos National Laboratory. If Rasmussen's project works—if the "Los Alamos Bug" ever comes to life— it will redefine our position in the universe. The thing we call life will cease to be an anomaly.

Perhaps unsurprisingly, Rasmussen has been accused of playing God; there have even been suggestions his project should be halted. If he wants to dissipate any such concerns, all he has to do is list a couple of the ingredients of the Los Alamos Bug. His recipe for life will take a different path from that taken by the Pilbara microbes—and everything else on Earth. In fact, some would say the Los Alamos Bug isn't life, but a little ball of soap. Basically, it's a fleck of washing powder: soap plus a light-sensitive compound rather like the stuff that makes your shirt glow whiter than white. As Rasmussen wryly points out, you could buy the ingredients at your local grocery store. Hardly the stuff of sci-fi nightmares.

Soap molecules are based on triglyceride molecules—fat, essentially—but with distinct properties at each end of the molecule. One end loves water. The other hates it. Put them in water, and the molecules arrange themselves rather like the petals of a flower, with the water-loving ends facing out into the water, and the water-hating ends in the center. Oil and grease molecules get trapped at the center of each "flower" and are carried away from whatever it is they were clinging to.

The reason for choosing what is little more than a ball of fat (in fact, they are known as fatty acids) as a basis for the next generation of life is simple: it provides a useful container. In water it creates a neat, self-contained structure that sits happily in the test tube. All it needs now is some genetics.

The Los Alamos Bug's genetics don't involve DNA. Instead it has PNA. The *P* stands for *peptide*: a short chain of amino acids, the building blocks of proteins. PNA, like DNA, is composed of two intertwined strands of amino acids but is much simpler to make. It also doesn't carry any electrical charge, which means it will dissolve in the fat; PNA embeds itself in the oily drop that defines the Bug, and waits for the chance to replicate.

That chance comes when things get hot. Above a certain temperature, the double strands of PNA separate. This exposes small electrical charges on some parts of the acid.chain, and these charges are attracted toward the water. The chain itself, the backbone of the Bug's genetics, remains in the oily drop, but the electrical charges pull it to the edge. Here they encounter short acid chains, even shorter than the PNA, that Rasmussen and his team plan to leave floating around in the water—a kind of life support system. Some of them will bond to the "bases" of the exposed PNA strands; if a few are of the right kind, the PNA strand will find itself paired up into a new double strand. Its charges neutralized, it will dissolve back into the oily drop. As the temperature changes, the whole thing will happen again and again—the Bug's genetics will be constantly replicating—with a chance of interesting mutations at every turn.

Not that it's a done deal. Rasmussen's team has only got as far as having growth and division; there's no gene replication as yet. Nonetheless, Rasmussen is convinced that when it all works—and it is when, not if, he says—the Bug will be alive.

Well, sort of. He concedes that if you define *life* as "life as we are," as we know it, then it's not life. That would take many, many years, he says; a cell is an immensely complicated system, and we don't know the half of it yet. Rasmussen is convinced, though, that by all working definitions the Los Alamos Bug will be alive.

It will, for instance, have a rudimentary metabolism that makes it reproduce. Some of the short peptide chain feedstock floating in the water will have light-sensitive molecules attached to one end. These molecules will make the chain electrically neutral and thus fat soluble; the Bug will end up "ingesting" these peptide chains. When day breaks, however, light will break the light-sensitive molecules off. The chains will be left with a net electrical charge that will cause them to seek out the charge in the surrounding water—they will migrate to the surface membrane of the Bug. As the light levels increase, and more and more of these chains try to reach the surface, there simply won't be enough surface. The drop, Rasmussen says, will split in two. It will replicate. The way the whole thing is designed means that the PNA's electrical properties stop these feedstock molecules from becoming involved in the Bug's genetics, keeping the process of growth and genetic mutation nicely separate.

It is still a struggle to imagine that ball of fat as being alive, though. Indeed, the *Nature* editorial questioning the value of a definition of "life" also questioned whether any of the attempts to build organisms from scratch can really be regarded as "creating life." And, looking at some of the projects competing with Rasmussen's, we are tempted to answer no. Take Craig Venter's project, for example.

Though it is received wisdom that nothing good can come of a urinary tract infection, Venter, the man behind the private endeavor to decode the human genome, might disagree. Venter is also on life's case, and his project is attempting to elucidate the mysteries of life by working on a bacterium that makes it burn when you pee.

Mycoplasma genitalium was first discovered in someone's urine in the early 1980s; the patient was suffering from an affliction called nongonococcal urethritis. It turned out that the organism responsible, which lives in human genital tracts, has the smallest genome on the planet. Where humans have around 30,000 genes, *M. genitalium* has 517. Even then, around 300 of those seem to do nothing useful.

Venter headed the team that first sequenced its genome in 1995. The organism's relative simplicity inspired him to strip it down to its bare essentials and see what it really needs to survive. Once its genome has been reduced to the bare minimum, Venter will have an idea of what is required for life, he says. It will also provide a useful biofactory; he plans to insert other genes into the bacterium that could enable the organism to perform tasks like synthesizing insulin. That is undoubtedly why Venter is attempting to take the controversial step of patenting the *minimal genome*.

He has worked out the genes required for this minimal organism, synthesized them. The plan, at the time of writing, is to implant them in a bacterial cell that has had its own genome removed. He has already proved his team can carry out such a genome transplant in principle, so there is no technical hurdle remaining. Nevertheless, although it is vaunted as a step on the path to creating life, what Venter is creating is essentially a new species of bacterium rather than new life. David Deamer, a biophysicist at the University of California, Santa Cruz, goes even further. The creature Venter's team are trying to produce, he says, is really just a "radically engineered organism."

The same can be said for an effort under way in Rome, under the leadership of Pier Luigi Luisi. Luisi's "minimal cell project" starts with a vesicle, a kind of container used for transporting stuff around within cells, and will add various chemicals and components until something like a full working cell appears. At Harvard, Jack Szostak is also planning to fill a vesicle with biological material, this time to see when its starts replicating. Szostak is happy to admit it's a long-term project with no definite end in sight; he's been saying proper artificial replication is ten to twenty years off for ten to twenty years now.

Even if Venter's eviscerated cells or Rasmussen's ball of fatty acids in a test tube end up "alive," that doesn't necessarily tell us anything about this thing we call life. So where do we stand? Christian de Duve, who was educated by Jesuits, talks of a *cosmic imperative*, where life arises (when conditions are right) as an inevitable consequence of the laws of physics. That's essentially what Rasmussen says too: that life is just a very efficient way of processing energy. The trouble with this view is that it still leaves us without a clear idea of what life is and what made it appear on Earth. Rasmussen counters this by arguing that the individual element and the overarching phenomenon are two different things; looking at a car doesn't tell us anything about traffic jams, he points out.

And there, perhaps, is where the anomaly of life leads us to a scientific revolution. If reductionism is a dead end, maybe we should turn around and head off in the opposite direction.

IN August 1972 the Bell Labs physicist and Nobel laureate Philip Anderson published an essay in the journal *Science*. Anderson has always been a provocative voice, and never more so than in this piece. It was titled "More Is Different," and it makes inspiring reading.

Drawing on his experience of science as a process, Anderson forcefully makes the point that the behavior of large and complex groups of particles cannot be understood by applying our knowledge of the properties of a few particles. In other words, as with the difference between cars and traffic jams, More Is Different. This, he asserts, is a real principle, not merely an observation. At each new level of complexity, "entirely new properties appear, and the understanding of the new behaviors requires research which I think is as fundamental in its nature as any other."

If we are to understand the cosmos we live in, he says, we're going to have to abandon reductionism; the ability to reduce everything to simple fundamental laws does not necessarily give us the ability to start from those laws and reconstruct the universe. "In fact, the more the elementary particle physicists tell us about the nature of the fundamental laws, the less relevance they seem to have to the very real problems of the rest of science."

The thing is, we are used to breaking things up to understand them: the lump of metal breaks down to atoms, the atoms break down to nuclei plus electrons, the nuclei break down to protons and neutrons, which in turn break down to quarks, and so on. That's how science has progressed over the last century, and what a success story it has been. Why would we change the methodology now?

Because otherwise we are not going to progress, is Anderson's retort. We are plagued by arrogant molecular biologists who "seem determined to try to reduce everything about the human organism to 'only' chemistry," Anderson says. "Surely there are more levels of organization between human ethology and DNA than there are between DNA and quantum electrodynamics." Each level, he suggests, might require a whole new conceptual structure.

Anderson concludes his argument with recourse to a historical dialogue:

F. Scott Fitzgerald: "The rich are different from us."

Ernest Hemingway: "Yes, they have more money."

We all know that extreme wealth does not come with a rule book that dictates a strikingly different set of behavioral norms. And yet we have all seen the evidence that such behavioral differences do exist. Similarly, there is no way, Anderson says, to use the reductionist method to work out how and why certain phenomena have come into being; we must instead observe where these "emergent" behaviors arise and try to work out the principles that caused such emergence.

More than thirty years have passed, and still almost no one is listening. At the turn of the millennium, though, two more physicists took up Anderson's stance. The Nobel laureate Robert Laughlin and the distinguished physicist David Pines published a paper in the *Proceedings of the National Academy of Sciences*. Citing Anderson's cry that More Is Different, they declared that the central task of physics in our time "is no longer to write down the ultimate equations but rather to catalogue and understand emergent behaviour in its many guises, including potentially life itself."

The basic idea of emergence is that when a system is composed of many interacting parts, it will organize itself in ways that seem surprising; all the various interactions between the parts will lead to behaviors that look astonishingly complex. The chemist George Whitesides showed this by putting small iron ball bearings into a Petri dish, then putting a rotating bar magnet under the dish. The balls self-organize into concentric rings, each of which rotates. There are physical rules behind this behavior—having to do with magnetic interactions and the way each ball is affected by friction—but we could never hope to elucidate them. Perhaps, though, we could find the more general "organizing principles" behind the emergent behavior and take those as a set of rules to be consulted when analyzing similarly complex systems. The idea is that other complex and seemingly inexplicable phenomena, such as protein-folding and high-temperature superconductivity, might also be described by these rules: find one, and we might be able to unlock a rich seam of phenomena—including the enigma of life.

The people involved in this effort certainly talk a good game. According to the Santa Fe complexity theorist Stuart Kauffman, "organisms are not just tinkered-together contraptions, but expressions of deeper natural laws." To Laughlin, those deeper laws, the principles of organization, are the "true source of physical law, including perhaps the most fundamental laws we know."

In 1999 Laughlin and Pines established the Institute for Complex Adaptive Matter at the University of California. The idea was to bring scientists together to look at the various inexplicable "emergent phenomena" they have identified and try to work out the principles behind them. They must have been doing something right, because in 2004 the National Science Foundation began to fund the work.

The idea that a whole new branch of science is opening up is certainly inspiring and exciting; work out what makes those little balls form their rotating rings, and we might not merely solve the enigma of life but also discover the true nature of dark energy and from whence come the variations in alpha. The reality, however, remains somehow disappointing. As yet there have been no breakthroughs or insights that have changed our view of the universe. Neither is there any evidence that many scientists are abandoning the reductionist approach. We have no clue what the emergent laws might look like. That doesn't mean Anderson, Pines, Laughlin, and Kauffman are wrong, but it does mean the enigmas they might solve are likely to remain unsolved for a while yet.

LIFE, for now, stubbornly remains an anomaly: something unique, mysterious, and—put simply—"special." It's a situation that doesn't sit well with science. Most scientists, for good reason, don't want life to be known as something special, the result of a "vital spark" or, as the book of Genesis would have it, a mystical quickening due to the breath of God. Being somehow special doesn't fit with the overarching theme of science in the twenty-first century, a theme that makes a point of how insignificant we are. Carl Sagan perhaps said it best.

> We live on a hunk of rock and metal that circles a humdrum star that is one of 400 billion other stars that make up the Milky Way Galaxy which is one of billions of other galaxies which make up a universe which may be one of a very large number, perhaps an infinite number, of other universes. That is a perspective on human life and our culture that is well worth pondering.

As the author George Johnson put it, we have learned to "revel in our insignificance." At present, however, the anomaly of life spoils our revelry a little. So, while we wait to see if we can explain life, or at least re-create it from scratch to rob it of all mystery, what are we going to do about it?

One obvious answer is to find it elsewhere in the solar system. Perhaps we are finding life so difficult to make because it isn't as obvious a process as Rasmussen, Venter, and company would like to imagine; perhaps life got established so quickly on Earth not because it is straightforward but because it arrived, ready-formed, from outer space. Though that would make us the descendants of aliens, this is not a particularly contentious idea, scientifically speaking. In the early 1990s NASA funded a study into what happens when a rock hits Mars, Venus, or Mercury. The study took several years, using a few desktop computers to simulate the trajectory of rocks as they were thrown out into space; it was eventually published in *Science* in 1996. The result was clear: planets and moons of the inner solar system must have been trading rocks for billions of years. The researchers showed that, because of the way Earth's gravitational field attracts debris, about 4 percent of the stuff flung off the surface of Mars will land on our planet.

That certainly fits with the facts. Dozens of meteorites found in the pristine preservative environment of the Antarctic ice fields have a geology that says they came from Mars. And if the rocks have been coming from Mars since the era when Mars was wet and well suited to growing life—an era that came before Earth was habitable—why would we doubt that Martian life could have hitched a rather sudden (and unsolicited) ride to our planet and started a branch right here?

The journey from Mars to Earth can take up to 15 million years—there's no guarantee of a direct path—and would expose any traveling microbes to huge doses of radiation. But we know that terrestrial microbes can shut themselves down and survive millennia without respiring or metabolizing. What's more, the "extremophile" bacteria we have found in sulphurous springs, deep ocean vents, and radioactive debris show us we shouldn't underestimate what conditions a microbe can enjoy. The Earth is teeming with bacteria that can survive the harsh irradiation they would experience on the journey to Earth.

Given this information, it's hard to argue that life couldn't have come here from elsewhere in the solar system. So maybe life seems so strangely hard to make because we have no idea how it started; maybe Earth's conditions did not generate life but merely provided a good home. It is an especially attractive hypothesis when we have two more life-related anomalies to consider: a possible contact with alien intelligence and an experiment that seems to have discovered life on Mars.

6

VIKING

NASA scientists found evidence for life on
Mars. Then they changed their minds.

Any discussion about the origin of life, the nature of life, the inevitability of life, has to confront a set of experimental results gathered by Gilbert Levin in 1976. Thirty years on, they are still the subject of debate in the scientific literature.

Today, Levin's company, Spherix, has its headquarters in an anonymous suburban business park that is a forty-minute cab ride out of Washington, D.C. According to its Web site, Spherix has "managed some of the largest pharmaceutical launches and recalls in the industry" and oversees "one of the country's most advanced, affordable e-government solutions for state parks." Apparently, Spherix has processed nearly seven hundred thousand camping reservations for Indiana's state parks, mostly through a call center. Somehow, these accomplishments are less than dazzling when you know that the man in charge of this operation once used his expertise to investigate other worlds.

Not that Levin's origins were particularly glamorous. He started his career as a sanitary engineer; the thesis he wrote for his PhD, from Johns Hopkins University, is titled "Metabolic Uptake of Phosphorus by Sewage

Organisms." As unappealing a read as this seems, it set him on a path to the red planet. While working in the public health department of the District of Columbia, Levin invented a new way to detect the presence of microorganisms. His technique speeded up the process of testing samples by making the organisms breathe radioactive carbon that could be detected by a Geiger counter. It was the same technique that, when he was working for NASA, later allowed Levin to attempt the detection of extraterrestrial life.

When the results first came in from the Viking mission that carried his experiment, Carl Sagan, the face of cosmic exploration and the hero of every space-loving child in America, phoned to offer Levin his congratulations: Levin, he said, had made the first discovery of life beyond Earth. A couple of days later, to Levin's enormous disappointment, Sagan took his congratulations back; it had all been a mistake. Ten years passed before Levin found the courage to stand up for his results. And despite the toll of the passing years—he is now eighty-one—Levin is still insistent that he found life on Mars.

MARS is Earth's sister planet. It may be a frozen waste with a thin, wispy atmosphere, but at least it has something we can work with when hypothesizing over the existence of life on its surface. Venus's atmosphere has a crushing deep sea–like pressure; Mercury and Pluto have no atmosphere; and Jupiter, Saturn, Uranus, and Neptune don't even have a surface we could stand on. In comparison, Mars seems positively welcoming. People have even come up with ideas for "terraforming" Mars; there are ways we could transform it into a planet that is habitable for humans. While this idea was once science fiction, now NASA researchers are drawing up work schedules.

Terraforming Mars is the culmination of centuries of human fascination with the red planet. The Babylonians knew it as the "fire-star," an angry, bloodthirsty sky-god, and the ancient Chinese, the Aztecs, the Greeks, and the Romans all felt similarly. We became a little more dispassionate about the planet for a while when we invented telescopes; in the seventeenth century Galileo Galilei and Christopher Huygens brought it down off its pedestal and charted its astronomical properties. Then, toward the end of the nineteenth century, it became mystical again as Percival Lowell tried to convince the world the planet harbored an intelligent civilization.

As soon as the march of technology made it possible, probe after probe was sent to examine Mars at close quarters. By the end of 1964 the Soviet Union had launched six craft toward the red planet. None of them made it, however; some rocket scientists joke about the "curse of Mars" because less than half of the thirty-seven craft we have sent there in the last half century have succeeded in their missions. At the time of the first Viking launch, there had been only six fully successful Mars missions in twenty-one attempts. Viking 1 reached Mars orbit on June 19, 1976. The next challenge, the next attempt to sidestep the curse, was to land a probe on the surface.

The first Viking lander was meant to touch down on Independence Day, but there was no safe site at which to land. In Puerto Rico the one-thousand-foot dish of the Arecibo telescope, later to become the backdrop for the Hollywood adaptation of Carl Sagan's bestseller *Contact*, was scanning the Martian surface and showed the proposed landing site to be littered with enormous rocks. The lander eventually went down on July 20, landing on the Plains of Gold. Nineteen minutes later, its signal reached Earth. Everything was go.

If the navigation team had done their preparation well, so had the team that would look for signs of life. As the mission was being designed, the life-seeking experiments were selected, honed, and then picked apart to eliminate all possibility the scientists would be fooled. The researchers were under no illusions about the importance of the task; these experiments had the potential to revolutionize our view of ourselves. Find life on Mars, and our perspective would be altered, suddenly and forever.

The mission team, together with four NASA-appointed review committees, had agreed on what would constitute success. If any of the tests showed a positive result, a duplicate sample of Martian soil would be heated to 160 degrees Celsius, a temperature that would kill any microbes, then tested again. If that test came up negative, the researchers could safely assume they had detected life, not chemistry.

It was only afterward—after Gil Levin's experiment met the agreed criteria—that they changed their minds.

On the face of it, Levin's achievements are extraordinary. Detecting life in your city's sewage is one thing; detecting microbial life using a robot scientist on a planet 200 million miles away is quite another. But Levin's "Labeled Release" experiment, sixteen years in the making, performed almost without fault.

The experiment gained its name through the radioactive carbon it used to "label" the gas released by anything that metabolized it. To produce a culture of microorganisms, you generally put some into a soup of nutrients in a Petri dish; they feed on the nutrients and begin to multiply. Levin tweaked this idea in a very simple way: by adding radioactive isotopes to the nutrients. The metabolism of microorganisms means that they will release gas derived from whatever they've been feeding on. If they've been feeding on radioactive carbon, a Geiger counter above the gas should go crazy. The plan was simple: add radioactive nutrients to a soil sample containing microbes, and watch for a rising graph from the radiation detector. Then, if it works, heat the soil sample to 160 degrees Celsius, killing the microbes, and repeat. You can add all the radioactive nutrients you like, but you won't get radioactive gas. It worked for finding microbes in suspect water, and it worked when tested on Earth, using California soil. And then it worked on Mars.

It was July 30 when Levin saw the first graph showing that Martian soil is just like California soil. A day earlier, the robot arm on the Viking lander had scooped Mars dirt into a box that distributed a little of it among four chambers. Each one contained half a cubic centimeter of soil. The chambers were sealed, and for the next twenty-four hours, the radiation detector monitored the background radiation in the air above the soil. It was a flat line.

Then the nutrient went in. It was a microbe's perfect lunch—with an extra kick from a little radioactive carbon-14. Fifteen hours later, the flat line shot upward. Radioactive gas was filling the microbe chamber. To start with, the assembled scientists were startled by the similarity to Earthbound data; they had seen this signature hundreds of times in their tests. Then they got over their shock and had a party. Levin went out and bought some champagne. He even got himself a cigar. They printed the graph, then everyone on the team signed it. The big hit show of the time was *West Side Story*, and

Levin wrote the title of one of its songs—"Tonight!"—on the top of the printout.

Levin was the happiest man in the solar system, but his joy wasn't to last. As agreed, the Labeled Release team later carried out a control experiment, heating one of the soil samples to 160 degrees before adding the nutrient. The line stayed flat, making the initial indication of life a strong scientific result. The Labeled Release team had met the four criteria that NASA had agreed signaled the presence of life on the red planet. By that time, however, the results of another experiment were in. And that one said there simply couldn't be life on Mars.

The two Viking landers each carried apparatus for four experiments. The second, the "Pyrolitic Release" experiment, seemed to give a positive result. During a five-day test, organic molecules, the basis of biology, were created by *something* in collected Martian soil. The scientists' best guess was that some kind of algae was responsible.

The "Gas Exchange" experiment gave a negative. It mixed a scientist's version of chicken soup—a broth of nutrients—with Martian soil. Analyzing the gases given off, the researchers concluded the soil contained nothing that had thrived on the nutrients.

Gilbert Levin's Labeled Release experiment, on the other hand, gave positive indications of microbial activity. In a way, the fourth experiment, the Gas Chromatograph Mass Spectrometer, which would test the soil for organic—that is, carbon-based—compounds, held the casting vote. Which is a pity, because it didn't work properly.

The thinking behind the GCMS experiment was, if there were organisms on Mars, the soil would be littered with decaying bodies: assemblies of carbon molecules. The experiment would take soil samples from Mars, roast them, and analyze the gases given off. If there was any carbon present, the experiment would detect the presence of volatile carbon-based chemicals.

Unfortunately, the experiment had problems. They had started en route: while Viking 1 was cruising toward Mars, a test showed that one of the three ovens in the GCMS apparatus, used to heat soil samples so they would give off gases, wasn't working. Then, on Mars, it turned out that the indicator showing a soil sample had been successfully delivered to the second oven didn't work either. Two out of three ovens had failed. And that was before Levin's experiment had even run. After its successful run, with the outcome of the mission resting on the GCMS's result, Levin held his breath while the GCMS's third oven was fed a sample. Six Martian days after the sample failed to register in the second oven, the same thing happened again. Not wanting to risk heating an empty oven, they went through the emptying routine—just in case—and waited for the next soil dig to come around. That was seventeen Martian days later. There was still no indication of whether the sample had been delivered, but the GCMS team went ahead anyway. The only data that came from the instrument showed that the oven still contained microscopic traces of the cleaning solvent used by NASA engineers prior to launch.

The GCMS experiment was run four times in total. The Viking 2 attempts, housed in an identical lander that followed Viking 1, at least registered samples in the ovens. But no trace of organic material was detected in any of the four runs. And no organic material, in the team leaders' interpretation, meant no life.

Naively speaking, it is inconceivable that there are no organics on Mars. After all, even our sterile Moon is littered with carbon that arrives in meteorite impacts. The solution put forward by the Viking team leaders was that some chemical in the Martian surface must break up organic compounds. It would, they suggested, do the same to Levin's nutrient, explaining his "positive" signal. The chief suspect was hydrogen peroxide.

The thing is, hydrogen peroxide has never been found on Mars—despite at least four extensive searches in the atmosphere and on the Martian surface. What's more, Levin points out, it is stable to temperatures of more than 160 degrees Celsius (320 degrees Fahrenheit). If hydrogen peroxide in the soil was breaking up the nutrient and releasing radioactive gas, it would have continued to do so after the soil samples were baked.

Nevertheless, the hydrogen peroxide argument fit with the negative result from the GCMS experiment. Thirty years on, the argument would still benefit from someone actually finding some.

———

AT the risk of muddying the waters, it has to be said that the GCMS result was not the only problem for Levin's Labeled Release experiment. A further procedure, carried out by Levin and his coworker Pat Straat, gave a puzzling result during experiments with the second Viking lander.

The consensus that chemical processes could explain the negative GCMS results was growing among the mission team; the prevailing idea was that ultraviolet rays from the Sun would produce hydrogen peroxide in the soil, which would then destroy all organic matter. So Levin and Straat asked the team controlling the sampler arm to move a rock and dig into the soil underneath, where there would be no hydrogen peroxide. The resulting sample gave another positive result in the Labeled Release experiment, punching a hole in the hydrogen peroxide argument. It also demonstrated, however, that a lack of light was not a problem to Martian microbes; they could live happily under a rock. Unfortunately for Levin and Straat, they already had evidence to the contrary.

On Martian day thirty-six, the team had put a sample of Martian soil into the chamber of the Labeled Release experiment. When the nutrient went in, something in the soil reacted, releasing radioactive gas just as in all the previous experiments. Then the chamber was covered over and left alone for seven days.

After a week in the dark, the team injected some more nutrient. Every time they had done this with microbe-infested soil samples on Earth, the Geiger counter had registered another increase; the microbes had gulped down the second helping. On Mars, nothing happened.

On the positive side, as we have noted, this result again stands against the argument that some compound, probably hydrogen peroxide, was responsible for producing radioactive gas from the nutrient—a prolonged lack of light would not affect the chemical process. But neither does it make a lot of sense if biology was involved.

One of the strongest arguments against life existing on Mars has always been the harshness of the environment: low temperatures, a wispy thin atmosphere, and the lack of liquid water all count against the development of living organisms. Levin counters this by pointing to the many subsequent discoveries of *extremophile bacteria* on Earth. Microbes have been discovered thriving in some of the most inhospitable places on our planet: in the freezing wastes of Antarctica, in the violent and scalding water around deep ocean vents, in volcanic rock, even in radioactive waste. At the time of the Viking mission to Mars, the existence of life in such places was unthinkable, but now it seems quite reasonable that life could take hold in Martian soil. What doesn't seem reasonable, given the tenacity of Earth's extremophiles, is that the microbes had died during a week in the dark. The experiment with the second lander, where microbes were apparently thriving under rocks, stands against that.

One possible explanation is that the sample taken from normal, exposed soil contained microbes that needed light, but there are other organisms, living under rocks, that don't. In the end, all we can say is that it does muddy the waters.

WHATEVER the truth about the complex web of results, the weight of evidence against the detection of Martian life—the negative GCMS result coupled with the hydrogen peroxide argument—was deemed compelling enough for the mission leaders to conclude that they hadn't found life.

Levin still remembers the shock of sitting in the first press conference to announce the outcome of the Viking experiments. Jim Martin sat next to him, and together they reeled as their team leader, Harold Klein, made the official announcement. The Viking mission had found "no evidence" of life on Mars, Klein stated.

"When he said that," Levin recalls, "Jim Martin dug me in the ribs and said, 'Goddamnit, Gil, will you get up and tell them you detected life?' "

He didn't. He says he was cowed by his relatively junior status, and that he also wanted to be conservative; he "didn't want to be out of step with anyone else on the team." He maintained this silence for ten years, the first three of which he spent trying to find alternative explanations for his own results. It was during that time that John Milan Lavoie Jr. got in touch.

Lavoie was an MIT graduate student who had performed many of the tests on the Viking GCMS. He was embarrassed at the way the GCMS results had been appropriated to quell speculation over life on Mars; the instrument's readings, according to Lavoie, should be treated with extreme caution.

Lavoie told Levin that the MIT-built apparatus had repeatedly failed in tests before launch. When given a sample of Antarctic soil, it had failed to find any organic compounds. That news was particularly striking to Levin, because all the various Viking experiments had been given the same sample to test prior to their acceptance onto the mission. When Levin had tested the sample—it was known as Antarctic Soil #726—his Labeled Release experiment recorded a significant rise in radioactive carbon in the air above the sample: Antarctic Soil #726 seemed to contain life.

A few years later, one of the engineers on the GCMS project approached Levin with a story similar to Lavoie's. Arthur Lafleur had been brought onto the project to help it meet its mission deadline, and had coauthored the paper that reported the negative findings on Mars. But, he said, the machine really wasn't anywhere near as sensitive as it needed to be to refute Levin's results.

Levin and Lafleur published a paper together in 2000, exposing for the first time some of the preflight results from the GCMS experiment. It had repeatedly failed to find organic compounds that were present in samples. Antarctic soils contained ten thousand organisms per gram of soil, but even at concentrations of 3 billion organisms per gram, the GCMS would have failed to spot organic compounds. Martian soil can probably contain no more than 10 million organisms per gram. In short, they said, the GCMS "was unequal to its assigned task."

By then, ironically, this was not a controversial claim. In 1996, at a NASA press conference, the associate administrator of NASA, Wesley Huntress, had said the same. The press conference was to announce the possible discovery of the signature of life within Martian meteorite ALH84001 (the issue remains unresolved today). The rock had arrived on Earth thirteen thousand years ago; it was recovered from Alan Hills in Antarctica in December 1994, and NASA scientists had found what seemed to be fossilized microbes.

A journalist asked the obvious question: Had NASA changed its tune? If this rock says there was life on Mars, how come the Viking GCMS found no organic material? Easy, said Huntress. For starters, the rock is a hint at past life on Mars; it has nothing to say about the present. Second, the Viking landers landed in a desert in order to find a safe place to touch down, and that "kind of reduced the probability of finding organic material on the planet should it be there." And third, Huntress added, the GCMS simply wasn't sensitive enough to rule anything out.

In 2006 the final nail was driven into the coffin of the GCMS experiment when a team of twelve researchers, including NASA's Mars expert Chris McKay, published a paper on the experiment in the *Proceedings of the National Academy of Sciences*. The sensitivity of the GCMS experiment, it concluded, was several orders of magnitude lower than originally thought. "The question of whether organic compounds exist on the surface of the planet Mars was not conclusively answered by the organic analysis experiment carried out by the Viking Landers," the paper states.

AT the party to celebrate the tenth anniversary of the Viking probes, Gil Levin stood up and gave a talk about all the possible reasons the Labeled Release experiment could have gotten a false positive. He listed fifteen or so and demolished each one. At the end of his talk he told the audience it was more likely than not that Viking had detected life. The reaction was not favorable—Levin describes it as "close to an uproar." He was not invited to the thirtieth-anniversary celebrations.

So how does he go forward from here? With caution, it seems. It would be easy for Levin to call for a rerun of his experiment, but he's not prepared to do that. He is advocating a careful approach to the case for life on Mars. As committed as he is to the idea that his instruments found evidence for life, he is not blind to all other interpretations. Even when other scientists come out with new arguments or evidence in support of his Viking results, Levin's attitude is surprisingly conservative.

Joe Miller, for instance, a cell biologist at the University of Southern California in Los Angeles, thinks he has spotted circadian rhythms in the gas emission from the Viking Labeled Release data. According to Miller, whatever was chomping on the free radioactive lunch showed the kind of cyclic metabolism we have; the gas release was not constant but varied in a cyclic fashion with a cycle of 24.66 hours—the length of a Martian day. Such rhythms in metabolic emissions are commonplace on Earth, and the discovery seems to rule out the idea that reactions involving nonorganic compounds such as hydrogen peroxide were responsible for the gas release. In 2002 Miller declared himself "over 90 percent" certain the Viking landers had found life.

Levin is not convinced by Miller's analysis, however. He recruited a math professor from the University of Washington to take another look, and he didn't find any significant pattern in the emissions data. "We didn't think it looked so positive," he says. When an Italian research group started to say they had found circadian rhythms, they got a lukewarm reception from Levin, too. "We're not satisfied," he says.

Levin knows how he would like to resolve the issue: he has redesigned the Labeled Release experiment to use *chiral molecules* in the feedstock. Certain molecules—glucose is one—come in two different forms. Just as left and right hands look similar but are not identical, chiral molecules have a subtle "handedness." While it makes no difference to their chemistry, terrestrial organisms will process one of these chiralities, but not the other. Probe the gas emitted in the new Labeled Release test for chirality, and you'll see whether life is involved in the emission: if there is a massive mismatch between the chiralities, you'll know the emission is biological, not chemical in origin. Other scientists are keen on the idea: Wesley Huntress expressed an interest, and NASA's Chris McKay, the man who leads the plans to terraform Mars for human habitation, said he'd like to copropose the experiment for a future mission. But Levin is cautious even here; the idea is not without its flaws, he says. We don't know whether Mars life has chirality preferences, for example. "It's possible they are both metabolized equally," he points out.

For now, then, all we have is the thirty-year-old results of an experiment that took place on the alien world 200 million miles away.

FOR some, the Viking mission is all in the past; there is simply no point in discussing it any further. Huntress, for example, who is now the director of the Carnegie Institutes of Washington, D.C.'s Geophysical Laboratory, still has a lot of respect for Levin. The problem, he says, is that astrobiology has changed since 1976. Any discussion of the Viking results has been rendered almost meaningless by the ongoing struggle to define what life is, and the conditions it needs to arise or survive—especially in light of the newly discovered extremophile bacteria.

Robert Hazen, an expert on the evolution of life who works upstairs from Huntress, offers a similar perspective: no one can agree on what a good detection of life would look like, he says. What's more, the life specialists are no longer so involved; after Viking, the biologists all left the field.

The void, it seems, was filled by geologists and atmospheric scientists. Almost everything in NASA's armory since Viking has been about detecting what we think are the *conditions* for life—at least life as we know it. Instead of looking for life, we are obsessed with finding out about the composition of the surface of Mars, looking at the rocks, and the patterns they contain that might or might not indicate the past or present existence of water. As you scroll through NASA's list of missions to Mars, it becomes clear that the biologists had their one chance with Viking and failed. The missions are now the preserve of other disciplines; before Viking and since, it has all been about rocks and weather.

The Mars Observer, launched in 1992 and lost before it entered orbit, "was designed to study the geology, geophysics and climate of Mars." In 1996 Pathfinder took photos, charted the weather, and carried out chemical analyses of rock and soil. Mars Climate Orbiter, lost on arrival on September 23, 1999, was designed to function as an interplanetary weather satellite. Mars Polar Lander was meant to dig for water, though it was lost on arrival on December 3, 1999. The Mars Global Surveyor has been monitoring the Martian surface, atmosphere, and weather, and investigating the composition of the planet's interior since September 1997.

Then, in 2004, came NASA's "robotic geologists," Spirit and Opportunity. The Mars Odyssey spacecraft continues to send us information about Martian geology, climate, and mineralogy. Mars Express is now searching for subsurface water from orbit (the mission's lander, Beagle 2, was lost on impact but would at least have looked for organic molecules). The Mars Reconaissance Orbiter is providing "an astoundingly detailed view of the geology and structure of Mars." At the time of this writing, Phoenix is on its way to the red planet. It will look for water ice and organic molecules.

Looking for life on Mars was a blip, a onetime opportunity, it seems. By almost every reasonable measure, we found it, but haven't looked again. Al-though almost no one doubts life *could* have existed on Mars in the past, and many experts think there *is* life on Mars now, it is Carl Sagan's conclusion—the possibility that we actually detected life on Mars is "vanishingly small," to borrow his phrase—that stands as the scientific consensus. And so the geologists can poke around with Mars robots, worrying about rock formations and liquid water, and managing not to draw a conclusion. There's no one who wants to stick their neck out like Levin did. And no one has to.

IF not a scandal, it seems a shame. This overwhelming caution, this softly-softly approach to looking for life beyond Earth, is postponing a glorious moment in the story of humanity. Peter Ward, a professor of biology, earth and space sciences, and astronomy at the University of Washington in Seattle, wrote a marvelous book about NASA's attempts to find (and create) life. In *Life as We Do Not Know It*, Ward is unequivocal about the importance of the quest to discover alien life. "The discovery of life beyond Earth would be monumental," he says. So why aren't we looking for life, not just tiptoeing around it? Apart from budgetary prudence and a sense that the last people who did that got their fingers burned, there is no obvious answer. It's not like we will find signs of microbial life beyond Earth and then stop looking for anything more. There's an even more important path to follow once we have made the discovery.

According to Martin Rees, the English astronomer royal and president of the Royal Society, "the prime exploratory challenge of the next fifty years is neither in the physical sciences nor in (terrestrial) biology. It is surely to seek firm evidence for, or against, the existence of extraterrestrial intelligence." Rees made the statement in a book laying out what twenty-five distinguished scientists consider to be the most important paths for science in the next fifty years. Elsewhere he argued that if he were an American scientist testifying to Congress, he "would be happier requesting a few million dollars for SETI [the search for extraterrestrial intelligence] than seeking funds for conventional space projects or particle accelerators." To Rees, the most distinguished scientist in Britain, and an international tour de force in astronomy, it really is that important.

What's more, it is not a fool's errand. Piet Hut of the Institute for Advanced Study in Princeton, New Jersey, has offered fifty-fifty odds of discovering intelligent aliens "out there" in the next fifty years. Hut knows it's a reasonable bet because we already understand that where there's life, intelligence will surely follow. In 2003 the Cambridge University paleobiologist Simon Conway Morris published a book called *Life's Solution*. In it he argues that, in order to survive in the habitats available to it, life must diversify and evolve solutions to the problems it faces. Life's solutions are constrained by the laws of physics, so although it might seem that there are innumerable solutions, there aren't; really, there are just a few. Which means that, wherever it evolves in the universe, life will look roughly the same. The chemicals involved might change, but the structures and machinery will necessarily converge toward a small set of possibilities. And this convergence, Conway Morris argues, will always—given time—lead to the evolution of intelligence because intelligence is one of the best survival tools available.

Once intelligence has evolved, an ability to use language to communicate confers a further advantage in the quest for survival, Conway Morris points out. And so the idea that distant worlds might be populated by intelligent beings able to communicate with each other, and eventually with civilizations alien to their own, isn't implausible. Indeed, if our next anomaly is anything to go by, Piet Hut may already have won his bet.

7

THE WOW! SIGNAL

Has ET already been in touch?

Science has a kind of golden rule, a principle that helps researchers distinguish between possible explanations for a phenomenon. The principle is called Occam's razor, and it says that, given a number of options, you should always go for the simplest, most straightforward one. If we apply Occam's razor to the signal received by the Ohio State University's Big Ear telescope in August 1977, we can conclude that it was a signal from an alien civilization. Why? Because it was exactly what we had been told to look for.

In September 1959, sandwiched between an article on the electronic prediction of swarming in bees and one on X-ray-induced metabolic changes in erythrocytes, the first scientific article on the likely characteristics of an alien communication was published in the journal *Nature*. The article was written by Giuseppe Cocconi and Philip Morrison, two physicists from Cornell University in New York. Cocconi had an unremarkable background, but Morrison's was more interesting. He earned his PhD under J. Robert Oppenheimer and played a vital role in the Los Alamos Manhattan Project. He was part of the team that traveled to Tinian Island in the West Pacific to assemble the atomic bomb that destroyed Nagasaki. After surveying the de-

struction, Morrison became a tireless champion of nuclear nonprolifera-
tion. He also helped found SETI, the search for extraterrestrial intelligence.

Morrison and Cocconi's paper in *Nature* suggested that anyone wanting
to attract another intelligent civilization's attention would use radio frequency radiation. It is relatively cheap and easy to produce, and it travels a
long, long way with a small power input. When it came to selecting a transmission frequency, they would choose one that spoke of some universal
number in the cosmos. Morrison and Cocconi's best guess was that an alien
civilization would use something associated with the most common element
in the universe: hydrogen. Any beings capable of communication would already have worked out and noted that hydrogen emits radiation at 1420
Mhz; this would be a number that would have special resonance everywhere
in the universe.

An alien signal, then, would come in at 1420 Mhz. And it would be, as
far as possible, only at 1420 Mhz. Sending a signal that is a composite of lots
of frequencies uses a lot of energy; anyone wanting to get maximum distance per kilowatt on their transmission will use a narrow frequency
range—a "narrowband" signal. As an added bonus, no natural phenomenon
emits narrowband radio frequency radiation, so the signal would make any
intelligent listener prick up their ears.

On August 15, 1977, an exact match for Morrison and Cocconi's signal
arrived in Delaware, Ohio.

IN the movie *Contact*, Jodie Foster gets a signal from space, and all hell
breaks loose. The U.S. National Security Agency tries to take over the project, the president is briefed, and his advisers descend on the scene in sleek
black military helicopters. Nothing like that happened at the Big Ear.
Around 11:16 p.m. Eastern Daylight Savings Time, the signal hit the first of
the Big Ear's two receivers. The telescope's computer recorded the signal's
arrival, a rise and fall in electrical current induced in the receiver's wire mesh
by an electromagnetic wave, then carried on recording whatever else came
in from the sky—nothing but noise, as it turned out. Three minutes later,
when the Earth had turned and brought the telescope's second receiver
around to stare at that same point in the heavens, the signal had gone.

A few hours later—by coincidence, it should be emphatically noted—Elvis Presley died. It was only three days later, while more than twenty thousand people filed past Elvis's open casket in Graceland, that the technician arrived at the Big Ear to stop the computer, print out the data, and wipe the hard disk clean. He came every few days; it was 1977, and the hard disk could only hold one megabyte. Perpetual data storage would be an unconscionable luxury for this long-shot project. On his way back up to Columbus, the technician dropped off the printout at Jerry Ehman's house.

Ehman, the man who spotted our best candidate for an extraterrestrial signal, is practically a legend. Other people would have spotted it too, he points out, with his typical modesty. But who else would have had the naive enthusiasm, the passion to write "Wow!" in the margin? Other people might have marked the printout with an asterisk or an arrow. Jerry Ehman wrote the exclamation that properly captures the profundity of the moment.

Much to his surprise, the name stuck, but he shouldn't be surprised. *Wow!* is a good summation of the importance of detecting an alien signal. It may even be an understatement. Talk to almost any astronomer—in private—and he or she will tell you it's the biggest thing there is. We are pouring huge amounts of energy into the biological effort to understand where life came from, how it arose on planet Earth, because it matters to us; it is, perhaps, our deepest question. Really, it boils down to this: Are we special? The best summation has been attributed to the science fiction writer Arthur C. Clarke: "Sometimes I think we're alone in the universe, and sometimes I think we're not," he said. "In either case the idea is quite staggering."

Clarke is right. If we are alone, that's extraordinary. If we are not, that's even better. Were we to discover that we are one of many life-forms on a planet that is one of many inhabited worlds, we would have a new perspective on being human—on being alive, even. And if we discover that some of that life beyond Earth is intelligent, a whole new vista of possible human experience opens up before us. We might, for the first time, have meaningful communication with another species.

That, really, is why we are looking for life beyond Earth—or, more accurately, suitable conditions for life. As we have already seen, the Mars Rovers were looking not for life but for the signature that there is, or has been, liquid water on Mars. It's not just Mars, though; the same search for the signs of water is going on with the Huygens probe on Titan, Saturn's giant moon. Jupiter's moon, Europa, has also had its conditions analyzed and been declared a potential haven for life. And these planets and moons within our solar system are just the beginning; the possibilities for life range across a whole universe full of planets.

We are living at a time of extraordinary progress in finding *extrasolar planets*; we did not spot the first one until 1988, but by August 2007 there were 249 confirmed sightings. There are several ways to do it. One is to identify anomalies in a star's orbit, due to a planet's mass pulling on the star. Or you can look at the starlight and see if it has become polarized—if the orientation of its magnetic and electric fields has shifted—by passing through a gaseous planetary atmosphere. Perhaps you'll see a "lensing" effect where the planet's gravitational field warps space around it and thus alters the path of the star's light. Then there's the "transit" method, where a star dims ever so slightly because a planet has passed across its face.

These are only a few of the techniques; there are plenty more, and they are all bearing fruit. In fact, it has got to the point where, if you want to make the news, just discovering an extrasolar planet is not enough. These days, to grab the front page you have to find a planet in its star's *Goldilocks zone.*

As with the idea of a Goldilocks universe, the name comes from the conditions: in the Goldilocks zone, the temperature is neither too hot nor too cold, but just right for the stable existence of liquid water on the planet's surface. So far, we have only found a few planets that orbit within the Goldilocks zones of their stars. In May 2006, for example, scientists announced they had discovered three planets, each with a mass equivalent to Neptune's, orbiting a star about forty-one light-years away. The outermost of these was in the Goldilocks zone. The following April, researchers announced the discovery of Gliese 581c, a planet orbiting a star in the constellation Libra. It too lay in its star's Goldilocks zone.

Though we are making great progress with finding suitable extrasolar planets, when it comes to detecting alien life there's a problem: the planets are *so* far away. There is a chance we might see signatures of possible life, or at least suitable conditions for life, in the spectrum of radiation from their surfaces or atmospheres, but we have little more to go on. If there are dormant life-forms on their surface, we won't ever know for sure. Without some dramatic leap in our technological abilities, there is no way for us to send probes or people to extrasolar planets. What we really need, then, is for that life to get in touch with us. It has never happened, or at least not in a way that convinces everyone who looks at the evidence. But the Wow! Signal remains our most tantalizing—indeed our only—possibility.

JERRY Ehman was in his kitchen when he read the printout from Big Ear. He was sitting at the table, with three days of data in front of him.

On the printout, the signal came in as "6EQUJ5." The letters and numbers are, essentially, a measure of the intensity of the electromagnetic signal as it hit the receiver. Low power was recorded with numbers 0 to 9; as power got higher, the computer used letters: 10 was *A*, 11 was *B*, and so on. 6EQUJ5 was the signature of a signal that steadily grows in intensity, reaches a peak, then falls away again. The *U* was the highest power signal the telescope had ever seen. The signal's spread was astonishing too: less than 10kHz. That's somewhere around a millionth of the transmission frequency. By anyone's definition, it was a narrowband signal at 1420 Mhz. Ehman knew what Morrison and Cocconi had said about the likely shape of alien signals. This fit exactly.

6EQUJ5 came up early in the printout—Ehman marked it with that *Wow!* and went through the rest of the printout to see if it happened again. It didn't.

It was enough, though. Eighteen years before the Wow! Signal hit Earth, before SETI had even been conceived, two physicists had predicted what an alien communication would most probably look like, and their prediction looked uncannily like the signal Ehman saw. If you believe that science should progress through theoretical predictions that are followed up by confirming observations, the alien hypothesis is a slam dunk.

So where has ET been hiding? The signal came from a single point in the heavens. Immediately on recognizing the signal, Ehman and his boss, Robert Dixon, consulted their star maps to see what astronomical body might be emitting it. The signal came from the constellation of Sagittarius, also known as the Teapot. Just to the northwest of the globular cluster M55 (to the east of the Teapot's handle) to be exact. There was nothing there.

Although the signal's shape didn't look at all like it had been created by accident, the researchers also looked for satellites or spacecraft—or even aircraft—that might have emitted a signal or interfered with terrestrial signals, creating something that looked like the Wow! Signal. Not only were there no man-made objects that could do it, the signal was of a frequency that global governments agreed was banned from use. There was no good explanation.

Three decades later, there still isn't. And there's very little more that one can say. The Big Ear researchers never saw anything like the Wow! Signal again. They looked for it more than one hundred times. Nothing. All the subsequent printouts were bland numbers, signifying the stubborn absence of anything interesting coming to us from the deep reaches of the cosmos. Most of our searches for alien intelligence have been similarly long, dark, eventless efforts. Occasionally something interesting has spewed out of the telescopes, but it has always turned out to be a spurious reflection off a satellite or a spacecraft, or interference from some piece of cosmic rock.

Though many have tried, no one has ever come up with such an explanation for the Wow! Signal. The researchers at Big Ear have analyzed a wide variety of possibilities: satellite transmissions, the harmonic frequencies of ground-based radio transmitters reflected off space debris, aircraft signals, terrestrial TV or radio signals, and anything else they could think of. Nothing could explain the characteristics of the observed signal. The first time I had contact with Ehman, he told me he was "still waiting for a definitive explanation that makes sense." Not that he believes it was aliens; he doesn't like to "believe" anything. It's just that it's the only satisfying explanation—if a one-off contact with ET can be classed as something satisfying.

In fact, it's this, the singular nature of the signal, that is its Achilles' heel. In *Contact*, Jodie Foster recorded hours, days, even weeks of extraterrestrial messages. The Big Ear received just one. Even the second receiver that looked at the same spot in the sky three minutes later saw nothing.

That certainly makes it tempting to dismiss the signal. It must have been some flutter in the electronics, or a bubble exploding in the telescope's nitrogen cooling system, or . . . something. If it was ET, then he, she, or it didn't broadcast for long—surely any deliberately broadcast signal would last for longer than three minutes?

The problem with that theory is that there's no reason for the assumption. Worse, everybody searching for extraterrestrial intelligence knows that intelligent beings could quite feasibly send one signal out into space followed by absolutely nothing else. They know that because we have done it ourselves.

In 1974 NASA arranged for the Arecibo telescope to beam a message out toward M13, a star-studded galaxy that seemed a good candidate for our nearest extraterrestrial homestead. The message was a stream of binary digits that, if you put them together right (carefully placed prime numbers provided clues), showed a crappy Atari Pong-style picture of a person, a DNA double helix, and our solar system. Anyone in M13 who picks it up—which won't happen for about twenty-one thousand years—may well conclude there is intelligent life out here. They may even be able to pinpoint where it came from. For that civilization on M13 it is likely to be a momentous event—their first contact with intelligent aliens. However, if they are anything like us, M13's brightest skeptics will smugly point out that you can't draw definitive conclusions from just one signal, no matter how well crafted. As any intelligent civilization knows, a sample of one is useless, statistically speaking. If ET really wanted to get in touch, there'd be two signals, at least. Wouldn't there? What a thought: we might have messed up our first communication with our cosmic neighbors. So perhaps we can take comfort in the fact that they seem to have made the same mistake.

IF there is no way to make the Wow! Signal make sense, there is also no way to invoke the other golden rule of science: repeat the observation. Today, there is no publicly funded search for alien intelligence—and there is no Big Ear. In 1988 the telescope was dismantled to make way for a luxury golf course. John Kraus, Big Ear's designer, learned Ohio Wesleyan University had sold the ground out from beneath his beloved telescope on December 28, 1982. He called it a day of infamy. "Ohio Wesleyan betrayed my trust and sold the land out from under the 'Big Ear,' " he wrote in April 2004. "What other discoveries and measurements might have been made if the telescope had not been demolished?" The fact is, there had been nothing more than a gentleman's agreement between Ohio Wesleyan University and Ohio State University, whose faculty had built the telescope. The local papers raised an uproar, and the OWU president resigned shortly afterward. The astronomers got together and offered the developers four times the land's value. The protests and the efforts, ultimately, made no difference.

Money, greed, and ambition have continually thwarted the search for extraterrestrial intelligence. Somehow, it seems more open to attack than any other branch of science. Perhaps because, as such a long shot, it is so vulnerable to cheap shots.

The first really cheap shot against SETI was fired just six months after the Wow! Signal hit Earth. Senator William Proxmire was looking for another recipient for his infamous Golden Fleece Awards. He handed them out to government-funded projects that he considered a waste of taxpayers' money. It was a great PR campaign for Proxmire, giving the voters exactly what they were looking for at the end of a difficult decade, but it wasn't always easy to keep coming up with targets—especially when he had committed himself to issuing one a month.

NASA's turn came around in February 1978 "for proposing to spend $14 to $25 million over the next seven years to try to find intelligent life in outer space." Scientifically, there was never anything wrong with the idea. The badly titled (by today's snazzy science-PR standards) "Microwave Observing Program" (MOP) had the support of mainstream scientists, and it had a moderate annual budget of around $1.5 million; it was a sensible effort to use microwave receivers to look for anomalous signals from outer space. Nevertheless, Proxmire's attention made it vulnerable, and, in 1982, he went in for the kill, tabling a legislative amendment that cut all federal funding for MOP. Fortunately, Carl Sagan came to the rescue.

Sagan's influence can be measured in TV viewing figures. His series *Cosmos*, produced in 1979, was the most-watched public program in America until the 1990s. Around 600 million people have seen the series and gained Sagan's charismatic, inspiring, and breathtaking perspective on the universe. When, in 1982, Sagan met with Proxmire, he was at the height of his influence. Proxmire listened to Sagan's arguments in favor of SETI and backed down—he even apologized. Sagan followed up with a PR campaign of his own, backed up by a petition signed by some of the world's most respected scientists (with seven Nobel laureates among them), and cemented the search for extraterrestrial intelligence in the American mind as a worth-while—even a necessary—scientific endeavor. No wonder, then, that Nevada senator Richard Bryan refused to meet with SETI astronomers when he launched his attack on the program a decade later.

On October 6, 1992, the *New York Times* was enthralled by the prospect of a new extraterrestrial frontier for America.

ASTRONOMERS, moving beyond philosophical musings and science-fiction fantasy, are about to mount the first comprehensive, high-technology search for evidence of intelligent life elsewhere in the universe. The new search is scheduled to begin symbolically on Monday, the 500th anniversary of the day Columbus happened on the shores of America.

Almost exactly a year later, the same paper expressed a numb shock under the headline "ET, Don't Call Us, We'll Call You. Someday."

LAST year, on the 500th anniversary of Columbus's arrival, NASA announced a 10-year project to scan the skies for radio waves emitted by alien civilizations. As Columbus Day 1993 comes around, the program is being canceled, the $1 million a month needed to sustain it eliminated from the budget.

The writer George Johnson could not resist stretching the analogy.

It was as though the Great Navigator, having barely sailed beyond the Canary Islands, was yanked home by Queen Isabella, who decided that, on second thought, she'd rather keep her jewels.

This disaster for SETI was due to Bryan. He had tabled a late-night amendment to a bill that killed the funding. In support of his amendment, Bryan made the facile comment that "millions have been spent and we have yet to bag a single little green fellow. Not a single Martian has said take me to your leader, and not a single flying saucer has applied for FAA approval."

This time SETI's champions could do nothing. Seth Shostak, now director of the SETI Institute, the privately funded successor to NASA's SETI, recalls that they requested meetings with Senator Bryan, but Bryan wouldn't take them. Bryan's amendment went through, and the publicly funded effort to answer the biggest question on Earth was over. It never recovered; the *New York Times* registered its amazement at the shortsightedness of the move, but nothing changed. Public funding of SETI was finished.

At present, the money pot for SETI is provided almost exclusively by Silicon Valley entrepreneurs. When SETI lost its funding in 1993, Barney Oliver, the head of Hewlett-Packard's research and development division and the man who gave the world the pocket calculator, made some calls. Oliver's true love was not technology but astronomy and, in particular, SETI, and he got Bill Hewlett and David Packard to make a contribution to keep SETI's head above water.

It is entrepreneurs like Hewlett and Packard who, for reasons no one quite understands, have kept SETI alive to this day; their contributions have allowed SETI people to buy a little telescope time and to pay a few salaries. But Hewlett and Packard are now dead, and it is another of Oliver's contacts, Microsoft cofounder Paul Allen, who is the main source of funding. Nevertheless, the construction of the SETI Institute's very own telescope—the Allen Telescope Array—is stalling because Allen feels his contribution should be matched by public funds, and no one with control over a public purse is willing to give any money for the construction.

It's easy to see why people who are accountable for public money might shy away from funding a search for extraterrestrial intelligence. Jerry Ehman admits it's like looking for a needle in a haystack—"except that you don't know where the haystack is, and you don't even know for sure there's a needle in it." It's true that the search for intelligent aliens relies on a barrage of assumptions, and one has to hope that some of them are not too wrong. But the same could be said of the search for extrasolar planets—a venture that has no trouble getting public money.

Take the current vogue for finding planets within the Goldilocks zone. When we stop and think about our limited appreciation of what life might be like, and what conditions it can thrive under, that whole set of criteria based on the existence of liquid water stars to look pretty shaky.

Liquid water is not a necessary requirement for life to exist and flourish; in some circumstances it can be the kiss of death. Sulphuric acid might do the job for other forms of biology, for example; the atmosphere of Venus is rather like a cloud of battery acid, and scientists have speculated that its acid droplets could harbor life. That's precisely because there is no water around. It is water that makes sulphuric acid corrosive; in fact, the acid is a catalyst for the corrosion reaction, known as *hydrolysis*, where water splits protein molecules.

Similarly, engineers have found that some biological enzymes used in industrial chemistry work in the hydrocarbon fluid hexane as well as in water. There is even a chance that biology can work without carbon; its near-relation silicon can also act as the scaffold on which biological molecules are built. On Earth, water and carbon are abundant, and silicon is locked up in the planet's rocky crust—sand, for instance, is mostly silicon. It's no surprise, then, that terrestrial life is carbon- and water-based. On other worlds, however, the kinds of distant worlds we are straining to see, there might be a sandman staring back at us. And those silicon-based eyes might well have developed far from the Goldilocks zone.

If the development of sand- or sulphuric acid–based life broadens the criteria for the search for other habitats, it also makes SETI's job much harder; the communication is even more likely to be something we haven't considered possible. But just as it hasn't stopped the search for life-harboring extrasolar planets, neither does it render SETI pointless.

There have been attempts to do that. Perhaps most famous is the remark that the Italian physicist Enrico Fermi made in 1950: "Where is everybody?" Fermi's point was that, for all the vast reaches of space and the almost limitless possibilities for intelligent life to develop in the universe, we have not encountered any aliens or alien communications. Many answers have been raised to the Fermi Paradox, including suggestions that aliens might not want to visit or communicate with us, or that they are already living among us, but the most compelling explanations are that we are not really looking or listening, and if we were, we wouldn't necessarily know what to look or listen for.

It is certainly true that we don't know what a deliberate signal would look like. Morrison and Cocconi's idea seems to hold water but could be rather primitive. If an alien civilization is advanced enough to be beaming speculative signals into space on a regular basis, it's likely to be far more advanced than we are. To them, our ideas of what makes a good signal may be the equivalent of smoke signals or semaphore: hopelessly outmoded and inadequate.

Our best hope would be that the aliens communicate using a mathematical code—a string of prime numbers or the digits of pi, or some other cipher we believe to be a universal experience. But there are other options. A project at Harvard University uses spectra gathered from optical telescopes to search for signatures of "always on" laser light beamed from deep space. A Berkeley project is looking at 2,500 nearby stars for pulses of laser light that might have been emitted by a distant civilization. Most SETI projects, including the Allen Telescope Array when it is up and running, look for Morrison and Cocconi's narrowband radio signal; although it would bear no information (at least none that we could detect using the current generation of instruments), the repeated observation of such a signal might release enough funds for us to build radio telescopes that could decipher any signal contained within it. Or that's what the SETI Institute is hoping.

WHERE does all this leave the Wow! Signal? Inconclusive. The fact that it came from an empty region of space, not somewhere known to be a candidate for the development of alien life, means the best we can suggest is that it was a signal from an alien spacecraft, perhaps an identifier beacon aimed momentarily and erroneously in our direction as a civilization migrated through the cosmos. But here we stray into the realm of science fiction.

Interestingly, the SETI Institute Web site's take on the Wow! Signal invokes another anomaly. "You wouldn't believe cold fusion unless researchers other than the discoverers could duplicate it in their labs. The same is true of extraterrestrial signals: they are credible only when they can be found more than once." Don't take it at face value, it suggests, but do look for more examples.

Are we looking? Not really. The search for aliens is for enthusiasts only. Considering what scientists say is at stake, this ought to be a scandal. The Wow! Signal, if it is what it seems, is a classic Kuhnian anomaly: follow it up, and we could radically alter our understanding of the cosmos and our place in it. It would be Copernican in scale. And yet it is, effectively, ignored.

On the bright side, there is still hope for clarifying the nature of life and our place in its hierarchy—and it lies much closer to home. If Martin Rees had his way, and SETI were to be modestly funded, it would lead us to examine the farthest reaches of space for clues to the essential nature of life. But it turns out that another terrestrial anomaly could shed more light on the matter. This creature—if it can be called that—bridges the gap between living and nonliving matter in a way that has never been seen before, and analysis of its genetic code is rewriting the history of life on Earth.

It's quite an achievement for a humble virus.

8

A GIANT VIRUS

It's a freak that could rewrite
the story of life

Pity the poor souls responsible for drawing tourists to Bradford, Yorkshire. First there are the dark, satanic mills of the city's industrial past. Then there's the fact that the Yorkshire Ripper, a notorious serial killer of prostitutes, lived here. The Brontë sisters were born and lived part of their lives nearby, but their lives were hardly long or happy. Emily died from tuberculosis at thirty, the year after *Wuthering Heights* was published. Charlotte, the creator of *Jane Eyre,* died at thirty-nine in the early stages of pregnancy. Today, in the United Kingdom at least, the city is better known as the site of violent race riots in the summer of 2001.

And then there is what may turn out to be the city's most important contribution to science. In 1992 Timothy Rowbotham, a microbiologist with the United Kingdom's Public Health Laboratory Service, was charged with finding the root of a particularly nasty outbreak of pneumonia in Bradford. His detective work led him to sample the water at the base of a hospital cooling tower. When he took his samples back to the lab, he found they contained amoebae. That was unsurprising in itself, but the amoebae seemed to have been infected by something, some microbe, that he couldn't identify. Rowbotham named it *Bradford coccus*, perhaps one of the least glamorous epithets ever given. Not that Rowbotham cared. He had other things to do; he put the unidentified microbe in deep freeze and moved on to the next job.

Eleven years later we learned that Rowbotham had found a monster virus. It is by far the biggest virus known to science; it is huge, around thirty times bigger than the rhinovirus that gives you a common cold. And it is staggeringly hard to kill. Most viruses can be destroyed by high temperatures or strong alkalis, or shaken to pieces by sound waves—but not this one. That's not what has made scientists sit up and take notice, however. This giant virus's biggest impact won't be on the health-care systems of the globe. It will be on the history of life on Earth.

WE have only known about viruses for around a hundred years. Toward the end of the nineteenth century, Dimitri Ivanovski, a Russian biologist, was sent to find out what was blighting the Crimean tobacco crop. Whatever it was, it was getting through the porcelain filters the laboratory technicians were using to sift out bacteria. In 1892 Ivanovski published an article describing the new, minuscule kind of pathogen he had found. Martinus Beijerinck, a Dutch microbiologist, eventually gave the pathogen a suitable name in 1898: *virus*—a Latin word meaning a slimy liquid or poison.

Though the virus trail was blazed by two Europeans, it was an American who got the most recognition. In 1946 Wendell Meredith Stanley won a Nobel Prize after isolating the tobacco mosaic virus. Interestingly, Stanley's Nobel was for chemistry. Though they affect living systems, viruses have almost always been seen as merely chemical, not biological. In fact, they are viewed as almost mechanical: vicious, brutal, violent, powerful machines, hell-bent on reproducing themselves, but unable to achieve this on their own. Viruses can't exist without a living host to make proteins and energy for them. They are evolutionary aberrations whose existence necessitates destruction, rather like the cruelly amoral machines in the movie *The Terminator*. They are not part of the web of life.

There's just one problem with this traditional view, however, and it is sitting in a freezer in Marseille.

MARSEILLE, the oldest city in France, is now a world center for disease research. That expertise most likely arose because, when the city was founded by the Phoenicians in 600 BCE, and its harbor opened a gateway to the Mediterranean, North Africa, and the West Indies, it also opened a gateway to the plague: the first cases of bubonic plague arrived in Marseille in the year 543.

Plague is another example of the finely honed capabilities of the microorganism. Inside its flea-host, the plague bacteria multiply and block the entrance to the stomach. The flea can't be sated, no matter how much blood it sucks from its own host—usually a rodent—and so it feeds madly. The blood reaches the bacterial plug and is then vomited back up, laced with bacteria that infect the next thing the flea bites. And so it goes on. And on and on.

In 1346 a boat from the Middle East brought another plague to Marseille; the eventual European death toll was 25 million. We have short memories, though, and are motivated more by greed than by common sense. When, in 1720, a boat arrived in Marseille with several known cases of plague on board, the port authorities put it under quarantine, but the city's merchants wanted to trade its cargo of silk without delay. They put pressure on the authorities, who lifted the quarantine order. Thus began Marseille's Great Plague. Within two years, fifty thousand people had died in the city—more than half the population. Another fifty thousand died in the regions north of the city. No wonder the disease researchers in the medical faculty of Marseille's Université de la Méditerranée are among the finest in the world.

The president of the university is Didier Raoult. His biography reads like a list of things you're glad someone else knows about: he has degrees in bacteriology, virology, and parasitology. He has scraped out the teeth from plague victims; at the turn of the millennium, while the rest of us were planning the ultimate New Year's Eve party, he was picking out DNA from the teeth of exhumed fourteenth-century skeletons in order to test whether they were killed by a bacterial plague or a deadly Ebola-like virus. Raoult is passionate about pathogens. So when Timothy Rowbotham offered to send him a freeze-dried bacterium that defied all attempts at classification, of course he said yes. He couldn't have known then just what a mire he was stepping into.

First, the sample went under a microscope. Rowbotham was right; it certainly looked like a bacterium. Next, it passed the standard test for bacteria: the Gram stain. This is a series of chemical stains applied to a sample suspected to contain bacteria. It always comes up purple for bacteria and pink for anything else. Raoult's sample came up purple.

That's why Bernard La Scola, a bacteriologist in Raoult's group, took the next step and set out to classify exactly what kind of bacterium they were dealing with. This involves another standard routine that probes a molecule called *ribosomal RNA*, which helps the bacterium make proteins. Unfortunately, the sample didn't have the molecule in question. Nearly thirty searches later, La Scola still hadn't found it. So he took the cover off his electron microscope—a thousand times more powerful than his standard optical microscope—to have a closer look. And that is when he was confronted with a monster.

The bacterium was in fact not a bacterium. It was a giant virus. The team christened it "Mimi"; when they announced their discovery in *Science* in March 2003, the team said they chose the name because it is a mimic, closely resembling a bacterium. (Raoult subsequently admitted there is a less clinical side to the naming, however: his father used to make up stories centered on an amoeba called Mimi. Since the giant virus was first discovered inside an amoeba, to Raoult, it seemed sweetly appropriate.) The announcement took just one page; it simply said the French researchers had found the largest example of a *nucleocytoplasmic large DNA virus (NCLDV)*.

Biologists have a number of classifications for viruses. There is even a committee, the International Committee on Taxonomy of Viruses, that takes into account the viral properties in order to put it in the proper group. The committee considers issues such as the type of nucleic acid (RNA or DNA), the type of host, the shape of the capsid shell that encloses the genome, and so on. DNA viruses—herpes, smallpox, and varicella zoster, the virus that causes chickenpox and shingles, are examples—have a genome composed of DNA that sits within a protective protein coating. The NCLDV classification denotes the larger viruses in this group, and the Marseille giant virus is the largest of them all. Imagine standing next to a man who's the height of a twelve-story office building. That, to most other viruses, is what this freak looks like.

The view down Bernard La Scola's electron microscope shows Mimi—like all viruses—looks like some kind of crystal. It doesn't look baggy, like a cell or a bacterium. It looks like something that has arranged its structure according to neat architectural principles. Its head is an icosohedron, multi-faceted, like a well-cut gemstone. It looks well-organized, disciplined.

And it is. Unlike other viruses, it has a genome that is a model of restraint. Where most viruses have a headful of "junk" DNA that seems to serve no purpose, most of Mimivirus's genes perform well-defined tasks. And what tasks. There are genes, for example, that encode for the instructions and apparatus for making proteins. This violates biological dogma straightaway; viruses are supposed to let their hosts make the proteins. Some of the protein-making apparatus in Mimivirus is exactly the same as you'd find in all the things we call "alive." There are also genes for repairing and untangling DNA, for metabolizing sugars, and for protein folding—an essential step in the construction of life. The Marseille researchers found Mimivirus is the proud owner of a grand total of 1,262 genes. (The typical virus has 100 or so, but only uses around 10.) Scientists had never seen somewhere near half of them before, which has the Marseille researchers excited. However, it is the ones they *had* seen before that are causing the most fuss. To understand why, we have to go back to 1758, when Carl Linnaeus, a Swedish naturalist, published the tenth edition of his revolutionary book, *Systema natura*.

Linnaeus's volume did away with the simple but unenlightening system of his day for naming and grouping biological organisms. Instead, Linnaeus grouped organisms by their shared physical characteristics. In many ways he laid the groundwork for Charles Darwin; Darwin's theory of evolution by natural selection also examined why different organisms should share certain physical characteristics and arrived at the conclusion that if things look alike they are probably related in some way. Suddenly we had the notion of a tree of life, and we could start to think about tracing our ancestors.

Instead of everything having one (often very long) name, Linnaeus gave them two short ones. The first was its *genus*: *Homo*, for example. The second was the *species*, the subdivision that separates the members of a genus: *sapi-ens*, for example, or *erectus*. It was a neat system and is still biology's best. Though most of us are more familiar with *gray wolf* than *Canis lupus*, for some organisms Linnaeus's system provides the only familiar name: *Tyran-nosaurus rex*, for example, or *Escherichia* (better known as *E.*) *coli*.

The next classification revolution came in the 1970s, when Carl Woese looked beyond physical characteristics. Woese used the emerging technology of gene sequencing to allow grouping by shared characteristics in the genomes of various species. In doing so, he dared to redraw the tree of life.

At the beginning of that decade, life was thought to have had only two types of contenders. There were the *eukaryotes*, the advanced organisms like animals and plants whose large and complex cells contained a nucleus that held inheritable information. The other branch was the simpler *prokaryotes*, such as bacteria, which have cells without a nucleus.

In 1977, however, Woese published a paper that suggested the prokary-otes should split. He had been sequencing the genomes of various microor-ganisms, and something just didn't fit. A group of microbes called *archaea* were genetically distinct from bacteria; in fact they were genetically more like the eukaryotes. The archaea, which were characterized by living in high-temperature environments or emitting methane, might look similar to the bacteria, Woese said, but genetics said they represent a completely different evolutionary path. There were three kingdoms, not two. We now know ar-chaean organisms constitute a huge proportion of the planet's biomass—one estimate has it at 20 percent. Their signature is a seemingly inhospitable habitat. *Halobacterium*, for example, thrives in saline water. Others live in the intestines of cows, in hot sulphurous springs, in deep ocean trenches feeding off the black smoker vents, in petroleum reserves . . . the list goes on.

Woese's paper, published with his University of Illinois colleague George Fox, has an angry tone. It reads like a wake-up call to biologists; he speaks of the path to the tree of life being "obscured" by a narrow-minded scientific worldview. The words *prejudice, without evidence, taken for granted* come up. They talk of biologists' *predilection* for simplistic dichotomies: plant vs ani-mal; eukaryote vs prokaryote. But the biological world is not bipartite—eukaryote or prokaryote—the researchers announced. "Rather, it is (at least) tripartite."

That paper forcefully ushered in the age of the archaea as sitting alongside bacteria and eukaryotes like you and me. And that little parenthesized *at least* left the door open for more. Perhaps there are four branches, not three. Enter, Mimivirus—if it dare.

Despite Woese's calls for open-mindedness into the future, Mimivirus has not been welcomed with open arms. A virus that threatens to redraw the biological landscape again was never going to have an easy ride. And so far it hasn't. The jury is still out on whether Mimivirus should even be accepted as a form of life. This hedging seems extraordinary when Mimivirus is genetically more complex than some bacteria—all of which are considered to be alive. Why shouldn't Mimivirus be welcomed as a member of life's club? The only answer seems to be "because it is a virus." The orthodoxy says that viruses are parasites. Which means, logically, they can't have existed until after some other life-forms came into existence.

Logic is a wicked thing, though; it often relies on subtle assumptions. What if, for instance, viruses weren't always parasites? What if they evolved before life split into eukaryotes, bacteria, and archaea, but subsequently lost some of their independence? In that case they would have every right to be called alive—and they might hold clues, as many clues as the other three groups, about our *Last Universal Common Ancestor* (*LUCA*). Since LUCA is practically the holy grail of biology, it doesn't do to ignore the possibility, and the claim is not without foundation. Around half of Mimivirus's genes are unknown to science; no one has a clue what they encode. Considering how many genomes we have now sequenced, how many genes we have seen, that is rather surprising. Unless, that is, Mimivirus really is from another age. So perhaps in a bygone era Mimivirus wasn't a virus at all, but an independent, free-living organism that later fell on hard times and resorted to piracy. The 450 hitherto-unseen genes are one hint toward this; they may be relics of the distant past. But it is the seven genes it shares with every other living thing that provide the most intriguing clue.

Sequence your genome, and you'll find all kinds of interesting things. But among the genes that make you you, you'll also find sixty or so genes—the *universal core genome*—that link you to all of Earth's life. There are copies of these genes inside every biological cell on the planet, copies that write a textbook of the history of life on Earth.

We know this because genes, which are arrangements of acid molecules, are littered with mistakes: places where the acids have been linked up in the wrong order, or where something is missing altogether. This happens occasionally during the construction of a new copy; DNA is good at replicating itself, but it's not always perfect. Radiation can also cause mutations. Whatever the cause, the result is only occasionally disastrous; for the most part, the organism survives without any problem. These mutations then get passed down the generations and provide a hereditary characteristic. Just as it is possible to use certain physical attributes—a peculiarly beaked nose, for example—to pick out who is related to whom at a wedding, scientists can use the genetic mutations to work out the family relations in a group of organisms. If two of them have the same mutations in their core genes, they will have a common ancestry. By comparing all the various mutations, we are able to place organisms on an evolutionary tree.

Since Mimivirus has seven of these genes, Jean-Michel Claverie, another of the Marseille researchers, was able to compare its mutations with the known mutations in the rest of the living world and find its place on the tree. And it was rather a shocking discovery.

The team's 2003 *Science* paper had shown that analysis of the giant virus's proteins placed Mimivirus as a "deep branch" in the classification tree for the NCLDV viruses and left it at that. Less than two years later, they published the follow-up, again in *Science*, and this time they came out all guns blazing. The 2003 paper had taken just one page. Their November 2004 paper was seven pages long; Mimivirus was proving to be a gold mine. The complexity of its genome means that Mimivirus "significantly challenges our vision of viruses," the researchers wrote. They backed up their argument by referring to a 1998 paper that suggested a line of DNA viruses could have emerged before the three accepted domains of life split. The tree of life, they suggested, ought now to be redrawn.

Mimivirus, according to Claverie, occupies an entirely new branch, right down near the base of the tree. Its mutations suggest it evolved before the eukaryotes and their complex, structured cells—the very things it now infects. Most controversially of all, Mimivirus may even be directly responsible for the development of the well-organized cells that make you what you are.

———————

BIOLOGICALLY speaking, we eukaryote organisms are very impressive. Our cells have complex structure; somewhere along the evolutionary line the scraggy mess of the primordial cell turned into something with neat compartments and a nucleus that kept all our genetic information in one tidy package. The thing is, nobody knows how a cell first equipped itself with the extraordinary innovation that is a nucleus.

It was Franz Bauer, a celebrated biological artist (officially, "Botanick Painter to His Majesty"), who first described the nucleus in 1802, but in 1831 Robert Brown, the Scot who first observed Brownian motion, gave it the name that stuck. Since that time, biologists have come to appreciate just how astonishing the cell nucleus is; the complexity of its structure is matched only by the complexity of the tasks it carries out. Its DNA replication mechanisms, which create cellular life with consummate skill and ease, are the envy of every synthetic biologist.

The biologists do have a few ideas about how such a beautiful thing could have evolved. One respected possibility is that a merger between bacteria and archaea could have led to the formation of a nucleus; an archaeum trapped inside a bacterium provides the right kinds of conditions. This is fine, except that we also have evidence that cells with something like nuclei evolved before bacteria and archaea.

There are various other options; biologists can meet up and discuss them endlessly. It's just that they seem unable to decide which one is right. One of the few things they *can* decide on, though, is which, among all the options, is the long shot, the far-fetched idea that is allowed into the meeting only if it displays a badge marked *controversial*. Which idea is this? The virus, of course.

The champion for the virus idea for the origin of the nucleus is a Sydney-based microbiologist called Philip Bell. In 2001 Bell came up with a rather surprising hypothesis. What if a virus infected one of the scraggy, disorganized prokaryote cells and did something unexpected? What if, instead of just using the cell's molecular machinery to replicate itself and then move on, the virus actually took the reins? This new axis of evil, something somewhere between a bacterium and a virus, would have had abilities nothing else could match. And so, in evolutionary terms, it would have had a promising future. It would be able to engulf other organisms that had to make do with simple chemicals as food. Once it had engulfed them, the viral apparatus could simply take exactly what it needed from them.

There is circumstantial evidence that a virus—specifically, a DNA virus, Bell believes—could have been the first nucleus. Both are packaged DNA encased in a protein coating. In some relatively simple organisms, such as red algae, the nucleus can move between cells in a way that seems to reflect viral infection. Both package their DNA in linear chromosomes, while bacterial chromosomes are circular. The viral DNA strands even have primitive forms of *telomeres*, protective buffer zones at the end of the chromosome that are present in eukaryote chromosomes. (Their loss is thought to be linked to the process of aging—providing a link between viruses and the anomaly known as death, which we will explore in the next chapter.)

There are more similarities, but none is a smoking gun. Nevertheless, Bell has repeatedly stated that a DNA virus infecting a primitive archaeum could lead to something like a eukaryotic nucleus. The only flaw in that argument has always been that viruses are so unimpressive, so small, and so genetically uncomplicated. We know that cell nuclei are complex and impressive—how could a virus produce something like that?

For ten years, Bell searched for a virus that would be up to the task of becoming a nucleus. With the discovery of Mimivirus, he thinks he's found it; Mimivirus, he says, is the missing link. It's still a highly controversial view, however, because viruses have just not made it into the mainstream of evolutionary thinking. They were never considered alive, so how could they be part of the story of life? After all, viruses need something to host them, something to piggyback on. They are just *replicons*, bags of chemicals whose only purpose is to replicate themselves. And so the debate goes on. For the moment, for most biologists, Mimivirus remains an intriguing anomaly but nothing more.

A few biologists, though, insist their colleagues are in denial. Luis Villarreal, the director of the University of California, Irvine's Center for Virus Research, for example, sees viruses as "the world's leading source of genetic innovation" and thinks they are most probably the root of life on Earth. Much of the human genome, he points out, is viral in origin, so it is not a big stretch to imagine that LUCA, our Last Universal Common Ancestor, was some kind of virus.

The discovery of Mimivirus, with all its unexpected, unviral properties, has only served to cement Villarreal's view, and we have only just scratched the surface; there are probably plenty more giant viruses out there. In the last few years Craig Venter, the human genome pioneer, has been going back to life's roots, sailing the Earth's oceans, sampling the water every couple of hundred miles, and then sequencing the DNA in the bucket. Circumnavigating the globe in a one-hundred-foot boat called *Sorcerer II* is a wild way to do biology, and it has produced suitably stunning results. In the Sargasso Sea off Bermuda, Venter's team found more than eighteen hundred new species and more than 1.2 million new genes; so far, the trip has given us a tenfold hike in the number of known genes. And every bucketful of seawater—if you can call a two-hundred-liter container a bucket—contained millions of viruses never before seen by humans.

AS we have already hinted, the importance of getting to grips with viruses, rather than ignoring them, goes further than an abstract understanding of the tree of life. Viruses in general, and Mimivirus in particular, may hold the key to longer life, a key that seems rooted in their power to infect and commandeer a cell's machinery.

After Mimi was first identified in the Marseille laboratory, the researchers carried out various tests to determine the kinds of organisms it would infect. They ruled out human beings. Wrongly, it turns out. In fact, it is likely that a good many of us have antibodies to Mimivirus in our immune systems. When a research team in Canada examined a few hundred pneumonia patients, around 10 percent of them had antibodies to the virus; Mimivirus—or something like it—certainly used to infect humans. We already knew that many human incidences of pneumonia are due to unidentified microbes, and a study in France had shown that injecting mice with Mimivirus resulted in something like pneumonia. The final answer came when a technician in the Marseille lab came down with a fairly ordinary bout of pneumonia in December 2004. He was given a standard blood screening, which showed he had become infected with Mimivirus. The Marseille lab now operates with a slightly higher level of safety procedures, known officially as Biosafety Level 2.

Infection by viruses is almost universally seen as a problem. However, there are cases where it is potentially lifesaving. In 1988 Patrick Lee, then a professor on the medical faculty of the University of Calgary, announced in *Science* that a virus that is relatively harmless to humans can kill cancer cells. It is called a *reovirus*, and it seems to be drawn to cells showing abnormalities in a cell growth–regulating gene called *Ras*. Since most cancer cells have mutated Ras genes, it seems a plausible mechanism for fighting cancer without damaging normal cells.

Reovirus is currently being tested in clinical trials. The list of cancer cells it will kill is impressive—cancers of the breast, prostate, colon, ovary, and brain, and lymphoma and melanoma—but its power is not yet fully proven, and Lee and his colleagues are having to work hard to identify exactly what biological processes are involved in the viral action and reaction. The interesting thing is, the wider fight against cancer, an attempt to understand exactly the same issues, is now becoming closely linked with the fight against aging—and that is, in turn, causing us to reassess our understanding of just how eukaryote cells work. The prokaryotes don't age, so researchers are now going back to studying the detailed differences between eukaryotes and prokaryotes—which means revisiting the time when the tree of life started to branch. Since viruses like Mimi are now intimately involved in the debate over this era, it is just possible that Mimivirus has a deeper significance than anyone ever imagined. The origin of aging and death is linked to the emergence of the eukaryotes. And so is Mimivirus—especially if it really was, as a growing number of researchers now believe, the origin of the cell nucleus, the defining trait of the eukaryote cell. If there is a possibility that viruses can selectively infect and kill cancer cells, as Patrick Lee's initial findings show, perhaps that is because they go back to a time before the emergence of organisms whose cellular mechanisms go awry and cause them to age and die. It's an interesting speculation. However, as we will see in the next chapter, the possible role of a giant virus is just a small part of the anomaly we know as death.

9

DEATH

Evolution's problem with
self-destruction

In the summer of 1965, a young researcher from the University of Georgia caught a turtle in a Michigan marsh. It was a mature male Blanding's turtle, at least twenty-five years old. After noting its characteristics, he put it back. Thirty-three years later, in 1998, J. Whitfield Gibbons caught that turtle again. It was doing just fine.

Blanding's turtles are a biological enigma. The oldest known specimen was clocked at seventy-seven years old in the 1980s—a female that was still laying eggs. It's likely that, if she hasn't had her spine snapped by a passing truck, she is still reproducing now. Blanding's turtles don't get old and decrepit; they don't show any increased susceptibility to disease through their lifetimes. If anything, they get more vigorous with age; on average, the females lay more eggs every year.

Senescence, the deterioration with time that leads ultimately to death, is meant to be universal in the animal kingdom. According to the standard theory, everything gets old, falls apart, and dies. It's a good theory, but, in the light of the evidence, it doesn't add up—and it fails to add up in a very tantalizing way. The turtles are vertebrates and thus closely related to us in evolutionary terms. If our molecular machinery breaks down over time, so should theirs. But it doesn't. According to Caleb Finch, a professor of gerontology at the University of Southern California, the turtles are certainly "a sharp challenge" to the idea that our senescence is inevitable.

The turtles are not alone. Among the vertebrates there are several species of fish, amphibians, and reptiles that don't senesce. Finding out why they don't—and why we do—will have obvious immediate benefits. But it is a much more complicated story than anyone could have imagined. It's not really the Blanding's turtles that don't make sense. It's death itself that is the next anomaly.

WHY do living things die? Obviously, things kill each other—that's part of the natural order. But what causes "natural" death? It is a question that splits biologists. It has become like a game of Ping-Pong; over the years, theories have been batted back and forth as new evidence comes to light. Then, occasionally, someone steps in and ruins the game by pointing out that none of the theories fit all of the available evidence; we still have no winner.

One answer is that death is simply necessary—to avoid overcrowding, for instance. If nothing ages and dies, the biosphere is just going to start bursting at the seams. Even if each subsequent generation is stronger and fitter, survival will become ever harder as more and more organisms compete for the limited food resources. The best solution, then, is for the individual to sacrifice itself for the sake of the species. A simple piece of genetic programming that brings forth the next generation, then instigates self-destruction—or at least stops the repair process, allowing degradation to take its toll—is surely a sensible option, isn't it?

The nineteenth-century German biologist August Weismann thought so. He suggested that the body's resources can be categorized as either *germ* or *soma*. The germ carries the hereditary information, and its integrity must be maintained at whatever cost. Soma, which carries out the rest of the body's functioning, was "disposable"; once reproduction had occurred, the body would be wasting its resources if it put too much effort into repairing the havoc that time inevitably wreaked on the organism.

It sounds attractive, but it's a no-go. Evolution is supposed to select genes to benefit individuals and their offspring, not to benefit the group or species as a whole. If group selection works, evolution doesn't. In a famous rebuttal to group selection, the Oxford evolutionist Richard Dawkins dismissed it as "sheer, wanton, head-in-bag perversity."

In 1952 the British biologist Peter Medawar got around the problem. With great insight, he proposed a mechanism that would give a genetic selection for senescence. The power of natural selection is reduced as a creature gets older, Medawar pointed out, so a trait that gives an advantage before the creature has reached maturity (and entered into reproduction) will be selected for; a trait whose advantage only shows after the creature has ended its reproductive life will not. The converse is also true. A gene that disables you before you reach maturity will be (negatively) selected for; it will lower the chances of the organism passing on its genes. A gene that disables the organism much later in life will be, if not exactly selected for, at least able to survive into the next generation. And here, Medawar said, is the source of senescence. It's not about the inevitable ravages of time; it's the fact that late-blooming problematic mutations—genes for cellular machinery that breaks down late in life, for example—will be passed on, and will thus accumulate in a creature's genome. In humans, Huntington's and Alzheimer's disease provide examples of this process.

In 1957 George Williams expanded on Medawar's theme, introducing the idea of *antagonistic pleiotropy.* Pleiotropy occurs when a single gene influences more than one trait in an organism. Antagonistic pleiotropy occurs when that influence is advantageous on one trait while problematic on another. Medawar's effect could be achieved by a single gene that confers advantage—particularly reproductive advantage—when young but creates harm in the later stages of life. This quickly became the bedrock of the theory of aging.

Then, in 1977, Tom Kirkwood turned up to play. Kirkwood, a British mathematician, was unaware of Weismann's disposable soma idea when he lay in the bath contemplating the issue of aging (perhaps not an image one wants to dwell on). His idea, like Weismann's, was that aging was due to failures to repair somatic cells in the body. Kirkwood's insight was that those failures came from evolved traits that favored investment in reproduction.

This would manifest in the work—or lack of it—done by cellular machinery such as DNA repair genes and antioxidant enzymes on the somatic cells.

Kirkwood recalls his idea as being "highly controversial." That's because the prevailing view of the time, thanks to Medawar and Williams, was that aging is programmed. Over the years, though, evidence mounted up supporting Kirkwood's idea that aging is due to a slow, steady buildup of defects in our cells and organs. Gradually, programmed death fell out of favor. So much so, in fact, that when Thomas Johnson and David Friedman joined the Ping-Pong game by announcing they had found evidence of a genetic program for aging in 1988, some of their colleagues accused them of making up the whole ridiculous idea.

The pair were working at the University of California, Irvine, at the time. Their paper, published in the journal *Genetics*, showed that changing a single gene could make nematode worms live up to 65 percent longer than normal. Johnson and Friedman's paper went headfirst against the then-received wisdom that aging is the result of accumulated mutations in the genome. Apart from the snipes from colleagues, though, almost everyone ignored them. Until, that is, Cynthia Kenyon burst onto the scene and confirmed everything Johnson and Friedman had been saying.

Kenyon has close to celebrity status as a scientist. She is a molecular biologist at the University of California, San Francisco, and a founder and director of Elixir Pharmaceuticals, a company focused on "extending the quality and length of human life." Perhaps her most reported move was to put herself on a restricted diet as a result of her research: she stopped eating carbs like potatoes and pasta on the very day she discovered that the worms she was studying lived longer when there was no sugar added to their food.

Kenyon's initial breakthrough was not about *caloric restriction*, though. She had found another gene that increased a nematode worm's life span—and this time by 100 percent. The December 2, 1993, issue of *Nature* reported that *Caenorhabditis elegans* worms, which normally lived for two to three weeks, were living for up to six weeks. Worms that lived twice as long as they should seemed to tip the balance, and people started to discuss the possibility of a genetic switch for aging—and whether we could turn it off.

Since Kenyon's breakthrough, researchers have determined some of what makes the difference. The genetic tweak in the worms causes a cascade of molecular signals to go wrong. Those signals are similar to signals that the hormone insulin triggers in humans. Humans are hard to experiment on, though; it was when researchers discovered the signals are also similar to a hormone-driven cascade of signals in fruit flies that everything took off. Fruit flies have such a fast life cycle that they had already been co-opted as the workhorse of genetics research worldwide. The aging research worked too, and we can now use a genetic switch to lengthen fruit flies' lives. The same trick also works with bigger animals. We have a whole string of gene switches that we can flick to produce long-lived mammals—Methuselah mice, for example.

We still haven't got to extending human life spans, though, and for good reason. Our understanding of the processes of aging is still rudimentary, and no one is sure exactly what the trade-offs between longevity and ill health might be. Nevertheless, when you see what we *can* do for mice, it makes you wonder what we could do for humans. It's enough to give you, as the University of Michigan biologist Richard Miller puts it, "organism envy." No wonder many genetics researchers—Kenyon first among them—are now busy setting up companies whose aim is to find the elixir of life.

While those start-ups started up, however, a controversy was developing—and it is one that goes to the heart of the puzzle about aging and, ultimately, death.

In 2002 a large number of senescence researchers put their heads together and issued a "position statement." The group was headed up by Leonard Hayflick, one of the grand old men of gerontology, and the statement was signed by fifty-one scientists. Intended for public consumption, it warned against claims that misrepresented the science of aging and "victimized" those seduced by promises of eternal youth. "No genetic instructions are required to age animals," it said. "Survival beyond the reproductive years and, in some cases raising progeny to independence, is not favored by evolution . . . The processes of aging are not genetically programmed." In 2004, in the *Journal of Gerontology*, Hayflick opened an article with a blunt statement: "No intervention will slow, stop, or reverse the aging process in humans."

It contradicted everything the worm, fruit fly, and Methuselah mouse researchers were saying. How could Hayflick think, in the light of the published evidence, that you couldn't switch off aging? The answer lies in Hayflick's most celebrated discovery: *replicative senescence.*

IN October 1951 the research biologist George Gey went on national television in the United States and announced that a new era of medical research had just begun. He and his wife, Margaret, worked at Johns Hopkins University, where George was head of tissue culture research. The pair had spent the previous two decades searching for a human cell that would live forever in laboratory conditions; it would be the perfect tool with which to find a cure for cancer. When a thirty-one-year-old woman called Henrietta Lacks contracted cervical cancer and had a biopsy taken, the Geys found what they were looking for. George Gey faced the cameras and held up a vial containing cells cultured from Henrietta Lacks's cancer—the most robust and fastest-growing cells scientists had ever seen. "It is possible that, from a fundamental study such as this," he said, "we will be able to learn a way by which cancer can be completely wiped out."

Henrietta Lacks died from the cancer on the day that Gey went on TV. But, suddenly, cancer seemed like a prizefighter on the ropes, and huge resources were channeled into finishing off the fight. Lacks's legacy, the *HeLa* line of cells cultured from her cancer, have become another workhorse of biology. Her cells were instrumental in the development of the polio vaccine, they have been placed at atomic bomb test sites, and have even flown on the space shuttle. They continue to be used in biology labs worldwide, and their greatest achievement may be yet to come. In the fifty or so years that have passed since Henrietta Lacks's death, researchers have discovered many connections between cellular immortality, senescence, and the formation of tumors. What is perhaps the most important discovery came from the laboratory of Leonard Hayflick.

In the early 1960s Hayflick had been working toward understanding the mechanisms of cancer when he stumbled across the fact that normal cells could not be recultivated more than fifty or so times; in culture the populations would double for ten months, then suddenly die. Surprised but intrigued, Hayflick and his collaborator Paul Moorhead successfully repeated the process, then sent some samples to skeptical colleagues and told them when the populations would die. "Our predictions were met with disbelief, but when the telephone rang with the good news that the cultures had died when expected, we decided to publish," Hayflick later recalled.

The phenomenon Hayflick observed is known as *replicative senescence*. The truly intriguing thing about the process is that it has survived more than a billion years of evolution; it works in yeast in exactly the same way as it does in some human cells. Remove some of your fibroblast cells, for example, which are involved in creating the scaffolds on which new tissues grow, and you can culture them in a Petri dish. Then, suddenly, they just stop dividing and die.

Why should this be? It seems to be associated with damage to the DNA packed into the chromosomes of the cell nucleus. The counting mechanism, the ticking clock for senescence in our cells, is the *telomere*, a string of repetitive DNA sequences that cap the end of every chromosome. Telomeres stop the chromosomes from sticking together, but when the cell divides, the telomeres are not fully reproduced and become shorter on each division. Eventually, cells with enough depleted telomeres die. No one knows for sure how this mechanism progresses, but it has become central to the fight against cancer.

The tantalizing thing is, we know how to stop cells from dying. Cancer cells contain an enzyme called *telomerase* that restores the telomeres to their full length on each division. It is this that enables them to go into the runaway replication that causes tumors to grow so fast. We could avoid the shortening of our telomeres if our cells could produce telomerase. And they can.

In early 1998 a group of researchers led by Andrea Bodnar of the Geron Corporation in Menlo Park, California, announced they had put a gene that activates telomerase into normal human cells, and the cells had lived twice as long as untreated cells—and they were still going strong at the time of publication in *Science*. The cells looked good; they had the characteristics of young cells. The activated telomeres meant they had avoided the curse of replicative senescence; they were, to all intents and purposes, immortal.

The only problem is, you don't want immortal cells in your body because they would most likely grow into tumors. Telomere shortening might hasten our rate of aging, but it can also protect us from cancer. It's a trade-off. That is also true of another form of programmed cell death: *apoptosis*.

Apoptosis occurs in response to chemical signals. Viral infection, cellular damage, or just stress on the organism can stimulate these signals, which take the form of hormones, growth factors, and even nitrogen monoxide. All of them can tell a cell to die: enzymes called *caspases* start to break the cell down; the cell, effectively, eats itself. Apoptosis is an essential part of development—without it, your hand would not have separated fingers, for example—but when it goes wrong and allows cells to live forever, it also plays a role in the onset of cancer.

What we want to achieve in the fight against cancer is so much more complicated than simply having cells that live forever. Somewhere in here, though, is a tantalizing secret. "Perhaps," said the authors of a *Nature* review on cancer and aging in August 2007, "somewhere within the curse of the cancer cell's immortality there might also lie the secret of how we might understand and extend our own lifespan." Not that we should hold our breath for a cure; when it comes to understanding the roots of cancer and aging, "most of the fundamental questions remain unanswered," the authors admit.

SO we are left with two viable but contradictory theories. In the one camp, aging is controlled by a genetic switch that can only have arisen through some reproductive trade-off. In the other camp—Hayflick's camp—aging is simply the result of accumulated defects. Cells grow old and die because of copying errors and cell shutdown. It's not about reproduction or genes; it's about time.

Who's right? If we go by the scientific data, neither camp. There is evidence that contradicts both theories.

First, there is a fruit fly problem. When Michael Rose of the University of California, Irvine, began to breed long-lived fruit flies in 1980, their fertility declined. Things were looking good for antagonistic pleiotropy: long life came at a reproductive cost. But then, as the life span got even longer, the fertility began to rise—above the fertility of the normal, unenhanced flies. The flies were living 81 percent longer than the control group and were 20 percent more fertile. It's not the only time such an anomaly has been seen; Ken Spitze of the University of Miami bred fleas with increased life span and increased fertility. It shouldn't happen.

An additional problem for the theory comes from the observations of what caloric restriction—Cynthia Kenyon's diet of choice—achieves. Caloric restriction is thought to lower the metabolic rate and slow the production of cell-damaging chemicals known as *free radicals*. It certainly seems to lengthen life span—at least in mice, fish, worms, yeast, and rats. But the vulnerability to senescence that can be controlled through caloric restriction doesn't appear to have come about through antagonistic pleiotropy; controlling your calorie intake and thus lengthening your life span doesn't have the effect on fertility that it should. In experiments, female mice shut down their reproductive capability at 40 percent caloric restriction, but their longevity continues to rise if the restriction is continued up to starvation levels. Since resources are not being expended on reproduction beyond the 40 percent restriction mark, the extra longevity can only be coming from somewhere else.

Then there is the genetic switch problem. In research like Kenyon's, with *C. elegans*, single genes have been switched on or off to control aging. As her group point out in a 2003 paper in *Science*, in many cases there is simply no cost to this—not in health and not in fertility. Pleiotropy appears to be there—if you go so far as to remove the worms' reproductive systems, it makes them live another four times longer—but it is not a primary cause of senescence.

There's no recourse to the "grandmother gene" benefit either. While in higher animals such as birds and mammals a long postreproductive life would help in rearing the next generation, there's no need for it in the roundworm. They do not nurture their grandchildren, cooperate in groups, gather food for their young, or have to teach them how to fly. And yet *C. elegans* has a decent life span after reproduction. As the mathematician Joshua Mitteldorf puts it, "resources are being squandered on a useless life extension."

Mitteldorf, seeing the tensions between theory and experiment, became fascinated by the evolutionary biology of death. In 2004 he laid out all the evidence he could find in a paper published in *Evolutionary Ecology Research*. His conclusion was that there was no conclusion; the evolutionary origin of senescence remains a fundamental, unsolved problem.

Among the evidence, there is certainly no good news for the Hayflick camp, he says. If senescence were due to the accumulation of mutations over a life span, the older you take fruit flies and breed them for early mortality, the easier it should be to effect a change; damaging mutations should be there in spades. But the opposite is true. The older the flies are, the harder it is to breed for early death in the next generation. What's more, such a stubborn refusal to change is usually an indicator of a finely tuned mechanism that has been selected for by evolution. Death, here, is a program, and one that has been optimized.

Then there is the *mortality plateau*, which defeats all comers. The disposable soma camp says an organism won't repair itself after reproduction, so will be in continuous decline. The mutation accumulation theory expects the same result to occur by default (reproduction has nothing to do with it). The antagonistic pleiotropy theory is no different; the negative effects of the genes that gave advantage earlier in life will kick in one by one as the clock ticks onward. But culture a population of fruit flies, and the fraction that die per day only increases with their age until a certain point. After that, the fraction that die per day stays flat. That doesn't fit with any theory.

In other words, there's no good explanation for death. But if Mitteldorf has laid out the case against the popular theories of senescence with aplomb, what is he offering us in their place? The sheer, wanton perversity of group selection: species dying specifically to make room for the younger generation. Aging, Mittledorf says, evolved for its own sake, not as a by-product of better reproduction.

No one's buying that argument, though, because, as Mittledorf puts it himself, it "casts a shadow on a great body of evolutionary theory." He is right—and there is something familiar about this shadow. We are staring at the biological version of dark matter: a series of anomalous observations, complete with a possible explanation that opens up one can of worms too many. A seemingly good explanation would force us to rethink an ancient and vital part of the theory. Darwin's theory of natural selection, which *can-*

not work via group selection, is the biological version of Newton's universal law of gravitation. Does it need a tweak? It seems it might. Will the majority accept the tweak that has been suggested? Certainly not.

At the moment, we seem to be in the "ignore it" phase of this anomaly. The researchers looking into genetic switches for senescence have enough on their plates with finding the elixir of life. The other camp, those who think the former are selling (or at least researching) snake oil, have convinced themselves there is no anomaly. In April 2007 Hayflick published a paper under the title "Biological Aging Is No Longer an Unsolved Problem." Sweeping aside the ranks of senescence researchers who have exposed strikingly effective genetic pathways, Hayflick announced that the random accumulation of mutations is responsible for aging and death. If Cynthia Kenyon can make her worms live longer, that's because she is activating genetic switches that guard against certain diseases that would normally finish off a worm within a fortnight. She is mitigating against disease—admittedly, disease that is associated with old age—but she is not solving the problem of aging. Put simply, Hayflick and his followers believe the worms live longer because they are made stronger. And that's not the same as destroying time's power over biological molecules.

Kenyon and the other advocates of a genetic route to holding back the years don't agree and are aggressively pursuing their research. There are senescence switches, they say; find them, and flick them, and we can live forever. If only we could harvest the genetics of the long-lived species, the Blanding's turtle, say, or the Bowhead whale, which has an estimated life span of over two hundred years, we might find even more clues to immortality. But there are technical difficulties with doing this; culturing their cells is difficult, and there are legal issues with keeping and using such animals in research. And so it seems that the arguments about death will, like the Blanding's turtle, go on and on.

THERE is one clue that might take us forward. Cynthia Kenyon's studies of genetics tell us that aging is regulated by the same biochemical pathways in yeast, flies, worms, and mammals. If the mutations arose through random chance in the various species, each would have a different mechanism. But they don't; everything ages the same way. The reason is obvious, according to William R. Clark, a senescence researcher at the University of California, Los Angeles: senescence must have evolved in a common ancestor of today's species. Death, Clark believes, arose with the first eukaryotes, the organisms whose large and complex cells contain a nucleus that holds heritable information.

The story begins about 3 billion years ago, when the prokaryotes, the bacteria, and archaea, ruled the Earth. At some point these organisms evolved the ability to use light to split water into its constituent parts: the protons and electrons of hydrogen atoms, and oxygen. The protons and electrons made photosynthesis happen, giving the bacteria a very useful commodity: energy. The oxygen was released, an unwanted by-product of the process.

Most of the oxygen was absorbed by the green, iron-rich oceans of the era, creating heavy red particles of iron oxide that settled on the seafloor (a floor that has since been lifted out of the water by geologic shifts, the exposed red bands of rock giving us clues to this ancient past). When the iron was all used up, oxygen began to leak into the atmosphere above the oceans. As the concentration of oxygen in the air rose, it brought on the *oxygen catastrophe.*

Oxygen is highly toxic. When it breaks up, as it can in sunlight, the oxygen radicals formed can wreak havoc on biological cells. Around 2.4 billion years ago, the buildup of oxygen in the atmosphere eventually led to a mass extinction of the prokaryotes. They were, in effect, victims of their own innovation. Only those organisms living deep in the ocean, at a safe distance from strong sunlight, survived, evolving strategies such as aerobic respiration to cope with an oxygen-rich environment.

In fact, they did more than cope; they developed sophisticated and highly efficient means of turning oxygen into ATP, the fuel for all biological cells. It was such a successful innovation that it was soon pirated; as the eukaryotes emerged, they engulfed the energy-generating bacteria and put them to work. It was a doubly beneficial takeover because the bacteria had also evolved protection against the corrosive nature of oxygen, something that the eukaryotes took as part of the package.

There was just one problem for the eukaryotes: they had installed oxygen radical generators in the hearts of their cells. The mitochondria in our cells are the fossil remnants of the original ATP-generating bacteria, and though they allow us to generate energy, they also produce damaging oxygen radicals. There is, as they say, no such thing as a free lunch.

The problem, it seems, was big enough to require a truly innovative solution: sex. Or that's what Clark thinks. We still don't know exactly why sex evolved, but he is right; it may well have been provoked by the evolution of death. Sexual reproduction's process of gene swapping and shuffling allows DNA correction and repair, giving the descendant a potentially advantageous new set of genes. That is certainly beneficial in the context of the trade-off already going on between energy production and cell damage.

The only problem is, sex may have then encouraged more death mechanisms to evolve. If you have a new set of genes, you don't want the old, damaged ones getting in the way; if there is a means of removing the old set, it would prove useful. And such means exist. We know that in the group of aquatic organisms known as ciliates, a process of *apoptotic nuclear destruction* removes old DNA from the nucleus to make room for the new genetic combinations. It is a death mechanism, and it makes sense that it was positively selected for.

All because of sex. Which may well have evolved as a response to cell damage by oxygen radicals. Which, in turn, can be traced back to the mechanisms behind the production of the very energy that makes life worth living. Where there's life, it seems, death is close behind, but nobody has a full explanation for it. And then, somewhere in there, the sexual shuffling of genes has found a role.

The archaea and bacteria get by without sex and don't senesce. But when the first eukaryotes, our genetic ancestors, put these organisms to work to produce energy, it was with mixed results. They happily used the energy, which has enabled us to become all that we are, but it put the mechanisms of their eventual demise—death programs, if you will (and Hayflick certainly won't)—right into the heart of their cells. Only through sexual shuffling of genes could the cells mitigate against it.

If we haven't reached the true origin of death, is this at least the root of sexual reproduction? Was it just a repair mechanism designed for self-perpetuation that gained a life of its own and took an unexpected path? If this is the story, the survival of sexual reproduction as we see it today makes it an evolutionary spandrel, something that has arisen in the natural world as a by-product of another adaptation. And that might explain why it is that, as with death, we can't make sense of sex.

10

SEX

There are better ways to reproduce

In 1996 the arch-Darwinian Richard Dawkins published *Climbing Mount Improbable*, an outstanding exposition of the theory of natural selection. During his discussion of genetic mutation, and how it leads to advantage in the environment, he is forced to talk about the origin of sexual reproduction. "There are many theories of why sex exists," he says, "and none of them is knock-down convincing." Dawkins goes on to declare that he may at some time in the future summon up the courage to write a book about the evolution of sex.

He hasn't done it yet. In his 2004 book *The Ancestor's Tale*, he again admits defeat over the origin of sex. "To do justice to all the theories would take a book—it has already taken several . . . Yet no definitive verdict has emerged." In the end, he settles for discussing a consequence of sexual reproduction, rather than explaining its origin. The question of what is so good about sex is one that "better scientists than I have spent book after book failing to answer," Dawkins admits.

Dawkins is not alone in his frustrated silence at the prevalence of sexual reproduction. That chief among evolutionary biologists, the late John Maynard Smith, referred to an "evolutionary scandal" surrounding sex. Thanks to sex, said George Williams, there is "a kind of crisis at hand in evolutionary biology." In his book *What Evolution Is*, the biologist Ernst Mayr added his contribution. "Since 1880 the evolutionists have argued over the selective advantage of sexual reproduction," he says. "So far, no clear-cut winner has emerged from this controversy." Bringing things right up to date, a 2007 *Nature* review paper declared that "the explanation for why sex is so common as a reproductive strategy continues to resist understanding." You may never have thought too hard about it, but sex is a mystery.

The central enigma is simply that asexual reproduction, where an organism produces a copy of itself, is a much more efficient way to pass your genes down to the next generation. It does happen; many species, notably a number of reptiles and fish, perform limited amounts of asexual reproduction, copying themselves rather than collecting genetic material from a male (it is a female endeavor, producing only females). London Zoo houses a Komodo dragon that produced offspring without any male assistance in 2006, for example.

The puzzle is, why hasn't asexual reproduction taken over? Involve another organism by using sexual reproduction, and only half your genes get passed on. What's more, if a sexual and an asexual population are living side by side, every one of the asexuals is producing offspring while only half the sexual organisms are. Sex is a recipe for extinction; the asexuals will quickly take over the environment. So sex has what Maynard Smith called a "twofold cost": why would anything get involved in reproduction that is, genetically speaking, only half as effective as it could be—while also halving the speed of reproduction?

And that is just the genetics; we haven't yet mentioned the effort of competing for a mate, the inefficiencies inherent in the physical mixing of egg and sperm, and the problem of vulnerability to predators during the act of sexual reproduction. There's also the chance that the good gene combinations, the ones that evolution has selected for, will get pulled apart during the process of recombination and not get passed on. Almost every way a theorist looks at it, sexual reproduction is a disaster.

Countering this theoretical take, though, is the fact that, when you look around, sex obviously isn't a disaster; it is one of the most ubiquitous phenomena on the planet.

There is a quick and logical solution to this paradox. Evolution by natural selection is all about advantageous mutation; thus sex can only be so common because it confers a survival advantage. That advantage must come through the main outcome of sexual reproduction: offspring that are slightly different from the parent. And that difference must be valuable enough that it overcomes the enormous cost of using sexual rather than asexual reproduction.

Most observations of asexual reproduction show that it is an evolutionary dead end, a fast track to extinction. It comes and goes—lasting maybe a few tens of thousands of years—but it almost never persists in a species. It sometimes occurs in response to environmental stress, but it is not a universal strategy for most of the creatures that are capable of it. According to the orthodoxy, that's because any species that doesn't shuffle its genes can't survive natural mutations and shifting environmental conditions; in a variable environment there are obvious advantages to producing offspring that have different capabilities and tolerances.

In 2000, however, Harvard University's David Mark Welch and Matthew Meselson turned this argument upside down. They had been studying bdelloid rotifers, microscopic aquatic creatures that make great fish food. You can find rotifers almost everywhere there's water: in ponds, lakes, and roadside puddles, even in damp soil and mosses and lichens. What you won't find is a male bdelloid rotifer. These creatures reproduce without sex—and they have done so for longer than seems possible. Welch and Meselson's analysis showed they haven't needed males for eons; the 360 species of bdelloid rotifer have survived intact, using only asexual reproduction, for 70 million years.

It was this dogged survival, flouting biologists' best theories, that Maynard Smith called an "evolutionary scandal." It makes a mockery of the one argument in favor of sex: the idea that organisms need to shuffle their genes in order to survive in the long term. So although biologists see rotifers as the anomaly, it is really the rest of the natural world that needs explaining. Theory is all very well, but where is the evidence of the advantage of sexual reproduction? Just how catastrophic must the shifts in the environment be to make it worth paying the twofold cost of sex? To answer that, we have to look at what sex can do.

FIRST, let's consider the issue of the bad—biologists call them deleterious—mutations that accumulate through asexual reproduction. If an organism is just reproducing itself, any chance mutations in its DNA, caused by radiation damage, say, will be passed on. Thus, over the generations, the mutations will accumulate (the phenomenon is known as *Muller's ratchet*, after the discoverer of genome mutation through exposure to X-rays). The result is an organism that is always losing fitness. In sexual reproduction, on the other hand, there is always a chance that mutation-free blocks of genetic material will be transferred to the next generation.

It's a good, even obvious, theory, but the devil is in the details. The evidence in its favor is not nearly as positive as you might imagine.

Biologists gather such evidence—for and against—via some rather bizarre routes. William Rice and Adam Chippindale of the University of California, Santa Barbara, for example, converted a fruit fly from sexual reproduction to a cloning machine for their experiments. Aurora Nedelcu and her colleagues at the University of New Brunswick subject asexual algae to stress by heating to make their sexual reproduction turn on. (In the wild, it's the water temperature that operates this switch.) Matthew Goddard of the University of Auckland, New Zealand, performs genetic engineering on yeast cells, which can normally reproduce sexually and asexually, to switch off their sexual reproduction. Kellar Autumn of Lewis and Clark College, Portland, Oregon, made geckos run on treadmills, comparing the performance of those born through asexual reproduction and those born through sexual reproduction.

All these techniques—and there are more—are employed to test theories and see how sexual and asexual populations fare in different conditions. The answers, unfortunately, have not been as clearly confirming of the theories as anyone would like.

Autumn's asexual geckos, for example, were better athletes than the sexually reproductive ones, running farther and faster. But a previous study, carried out using a different species, found the converse was true. A series of experiments on water fleas found that asexual reproduction produced four times more deleterious mutations than sexual reproduction. But a study on nematode worms revealed absolutely no difference in the number of deleterious mutations in asexual versus sexual populations. Computer simulations of evolving genomes showed that the size of the population matters here too: small populations do better with sex, but larger populations of sexually reproducing species accumulated more deleterious mutations.

What about the idea that sexual populations can adapt to a changing environment more quickly because they are shuffling their genes? Again, the evidence is mixed. A 1997 study with yeast found no advantage for sexual species of yeast when adapting to a new environment. Another study, though, showed that sex can win out when the environment takes a turn for the worse, but the populations remain evenly matched if the environment improves. Yet another study, which took place in 2005, put a sexual and an asexual yeast strain into a test tube with minimal nutrients. The asexual strain won. When the same mix was smeared on a mouse brain, something supposed to mimic a highly varied environment, the sexual population won out. That result, though, contrasts with the findings of two Canadian researchers. In 1987 Graham Bell and Austin Burt showed that sexual reproduction *didn't* give the kind of genetic diversity that would profit an organism's offspring in a varied environment.

There is evidence, then, that sexual reproduction can increase the rate of adaptation in some situations, but it is hardly earth-shattering—and it is certainly not significant enough to account for the high cost of sex.

Further problems with sex arise when we look deeper into the mutations that are supposed to give sex an advantage. First, only a subsection of the virus family—RNA viruses—and the more evolved eukaryotes, such as humans, have high enough mutation rates to make it worth having sex to purge the deleterious mutations. Then there's the issue of *epistasis*, the interaction of genes. Multiple deleterious mutations in a genome can compound or diminish each other's effects, but the various studies that have been carried out into the effects of epistasis show no overall effects that would give sexual reproduction the edge.

Another possibility—and one that has been given a lot of credence—is William Hamilton's contention that sex is all about parasites.

Hamilton, who died in 2000, was an extraordinary figure. Not only for his academic prowess—one obituary called him "a good candidate for the title of most distinguished Darwinian since Darwin"—but also for his fearless personal exploits. He trekked through Rwanda at the height of the civil war, looking for ants (and was captured as a spy); he once jumped into the Amazon and used a thumb to plug a hole in his sinking boat; in Brazil he was knifed when he refused to yield in a street robbery. It was malaria, caught on an expedition into the Congo's jungle, that finally killed him.

Hamilton's imaginative approach to biology led him to coin a phrase that now resounds in the field: *the Red Queen hypothesis*. It was named after the character in Lewis Carroll's *Through the Looking Glass*; the Queen tells Alice, "here, you see, it takes all the running you can do, to keep in the same place." Hamilton used the idea as an illustration of the evolutionary arms race between an organism and its parasites. You evolve to get rid of your parasites; then they too evolve to use you as a host again. Sexual reproduction evolved as the best weapon in this never-ending struggle, Hamilton suggested.

Evidence in favor of this idea comes from various groups of researchers looking into the effects of parasites on yeast, beetles, sheep, and snails, among other creatures. Most show more successful reproduction and lower infestation by pathogens if their genes are reshuffled through sex rather than being replicated through asexual reproduction. With a variety of genetic makeups, it seems, there is a better chance that someone will live long enough to reproduce.

There is also evidence against the Red Queen hypothesis, however. Water fleas have shown no advantage over parasites when they use sexual reproduction. And the pesky rotifers don't fit within this paradigm, either. Why should they have managed to resist their pathogens for so many millions of years without sexual reproduction? There is evidence that, for rotifers, their advantage lies in genes that have adapted to help the organism survive in diverse conditions.

In 2004 Sarah Otto and Scott Nuismer struck another blow against the Red Queen. Their computer simulations of genetic interactions between a range of organisms in a large and varied environment—something like the real world, in other words—led to less sex, not more. So although the Red Queen hypothesis works in certain situations, it by no means accounts for the ubiquity of sexual reproduction. The only way it might work is if it is just part of a wide range of phenomena that, taken together, make sex a good reproductive choice. The Red Queen, they suggested in a paper published in *Science*, "might be less impotent with the right partner."

This seems to be the only answer left: that there is no one simple explanation for sex. Because none of the big, obvious explanations have panned out, the trend among researchers is now to look for a combination of smaller effects that give sex an advantage. One example is the way that sexual reproduction changes the genetic architecture. Experiments with artificial gene networks (more computer simulations) have shown that sexual reproduction gives rise to genomes that are "robust"; mutations don't have a strong effect on them. What is even more interesting, though, is the fact that sex also produces genomes that are more likely to be split into modules, self-contained entities whose genes have no effect outside the module. In sexual reproduction, the combinations of modules are shuffled rather than the genes, which reduces the risk of pleiotropic problems where one gene adversely affects another somewhere else on the genome. With a modular genome, the genes inside each prefabricated module are already tried and tested together and—if the creature has survived to reproduce—self-evidently do not produce enormously adverse effects (at least, not before reproductive age). Since the genes do not affect anything outside their own module, no amount of modular shuffling can produce further adverse effects, but there is still the possibility of advantageous recombination. Which means ongoing survival for the organism.

If it is true, it is still only part of the puzzle. In general, the random genetic drift due to chance variation offers the best hope of explaining the apparent advantage of sex. Research has shown that if populations aren't too large or small, and if the variations don't interact too much (that is, if pleiotropy is limited), sexual reproduction, more than asexual reproduction, can use genetic drift to enhance survival. But that's hardly a conclusive argument; biologists are still effectively offering up an argument that lacks strong supportive evidence. They just cannot answer the question of how we pay the twofold cost of sex.

TO Charles Darwin, the reason for the prevalence of sexual reproduction was "hidden in darkness." More than a century later, in 1976, Maynard Smith said the problem with sex was so intransigent it made him feel "some essential feature of the system is being overlooked." Three decades later, the problem is still here. It must be the longest-lasting scientific anomaly of them all. So, is it a Kuhnian anomaly?

It certainly has some of the hallmarks. In our efforts to combine a whole raft of small effects, it seems that our explanations for the origin of sex begin to look like something Kuhn called "a scandal": the Ptolemaic epicycles. These described the motions of the planets and stars as observed by the Greeks. The basic premise was that these objects revolved around the Earth. As observations got better and better, however, the astronomers had to repeatedly tweak their models of exactly how that revolving happened, adding layer upon layer of complication. It involved a gargantuan effort to keep the theory together—astronomy in those days largely consisted of anomaly-proofing the Ptolemaic system.

Early in the sixteenth century, an astronomer called Nicolaus Copernicus recognized that Ptolemaic astronomers had created a monster, and set about working out a better system. When he published *De Revolutionibus*, it all suddenly became clear. The motions of stars and planets made sense—and worked out ever so simply—if everything was in fact revolving around the Sun.

Is our theory of sex unwittingly Ptolemaic? And if so, can we see from whence its Copernican revolution might come?

Perhaps Maynard Smith's missing "essential feature" is the connection between sex and death (the subject of the previous chapter). If death—or at least cell senescence—is the root of sexual reproduction, the twofold cost of sex can plausibly be offset (perhaps more than offset) by the gain that comes with death: the ATP-generating machinery at the heart of every cell. Without it, we eukaryotes wouldn't have been able to take over the world. Let us run with this for a moment and see where it leads.

If sexual reproduction is a spandrel, a by-product of death, perhaps we can downgrade the primary assumption of biology: that the natural world is a fierce competition to pass on your own genes at everyone else's expense, using the best partner available (if a partner is necessary). Perhaps this drive is less intense than generally thought, and mitigated by other considerations, such as individual survival. If sex evolved as a result of the evolution of death in eukaryotes, survival must surely beat sex in the hierarchy of impulses. And we know that in most (but not all) sexually reproducing creatures, the desire to exist is stronger than the desire to reproduce.

Now let us imagine organisms living, as they ordinarily do, together in a group. (We are, necessarily, considering the higher animals here, but these are the creatures in which sexual reproduction is most firmly established.) They have a proclivity for sexual behavior and some impulse to reproduce, but also an awareness of the power of the group: their individual survival (the root of sex in our narrative) is linked in with the well-being of the group. What will happen?

There will be sexual behavior. As we well know, whatever the reason it has evolved, it has evolved to be a pleasurable bonding activity, at least in the higher animals. There will inevitably be reproduction. But there will also be consideration and effort directed at maintaining the integrity of the group so as to preserve the individual. John Maynard Smith once suggested that if the male partner contributes a significant amount to a sexual partnership, providing resources and working so hard that the female can produce twice as many offspring as an asexual female, the cost of sex disappears. Is it possible that a group dynamic such as that described above could more than offset the cost?

It is a difficult question to answer, but we can certainly make some interesting observations. Sexual creatures do often live in groups, and while it makes sense that each organism puts its own "best interests" at the top of its priority list, you can only work out what those best interests are when you take the whole group into consideration. It is not in a smaller male's best interest to try copulating with the sole female in the group, for example; if other males are much bigger, he could die in the attempt.

In some ways the issue parallels a well-known mathematical phenomenon known as *the stable marriage problem*. Imagine a party where a roomful of people are looking to hook up with a partner of the opposite sex. If all the men will only settle for the best-looking woman—and vice versa—almost everybody is going to end up unhappy. In 1962 two mathematicians worked out how, given a little compromise from everyone, you could actually make everyone happy. David Gale and Lloyd Shapley showed that if everyone compiles a ranking, in order of desirability, of potential partners, it is possible to arrange things into a stable equilibrium state. In this equilibrium, people are partnered in such a way that it is impossible to find a man and a woman from different couples who would both rather be married to each other than stay with their current partner. It's not the ideal for most individuals, but it is a satisfactory outcome for the group.

This is just one application of *game theory*, a mathematical tool used to track how the benefits and costs of decisions and actions will shape group behavior. Invented by the Hungarian mathematician John von Neumann, game theory has the central goal of finding an optimal solution to a problem, one where everybody involved in a situation is as happy as possible. Once this equilibrium is established, no one involved has any incentive to change it. The theory has proved a vital tool in a vast number of arenas: it helped establish the fragile peace of the Cold War; it has been brought to bear on economics and international relations; it explains how societies establish their social norms. In some ways, everything humans and animals do can be treated as a game. And that—according to Joan Roughgarden, at least—includes sexual reproduction.

Roughgarden is a professor of evolutionary biology at Stanford University and specializes in issues of sexual selection. In February 2006 she provoked an almighty row in the pages of the journal *Science* when, writing with two colleagues, she called for the wholesale replacement of Darwin's theory of sexual selection with a theory of *social* selection. The choice of sexual partner, she said, has to do not so much with reproduction, the propagation of genes, as with group bonding. And game theory, she said, shows why.

In her paper Roughgarden lays out a new theory to explain reproductive choices. Choosing the "best genes" is not involved in determining reproductive behavior, she says. Instead there is a kind of bartering system: opportunities to reproduce can be exchanged for services like attracting females, keeping a territory clean, or fighting off competitors.

Though many biologists have been critical of Roughgarden's ideas and approach, the theory does allow an organism to regain ground lost through sexual reproduction. She argues, for instance, that game theory shows social selection will increase the numbers of young raised to maturity. If group members are involved in performing the various functions necessary to group cohesion and survival, and these contributions mean that, in time, everybody gets a chance to reproduce because they are making a contribution, reproduction will be a more successful affair, pushing the numbers up.

It certainly provides a stark alternative to the traditional standpoint of biology—a standpoint that does have shortcomings. If you take the standard view of sexual selection, choosing a mate is meant to be a straightforward affair. It is based on the display of "good genes," usually manifest in the adornments and athleticism of the male of the species. For the most part the females choose (their eggs are limited; sperm is cheap and plentiful), and males slug it out for the chance to be chosen. However, recent studies showed that all that talk about females choosing males with the biggest antlers, or loudest roar, or, as in the case of peacocks, the most elegant tail feathers in order to get the "best genes" is just far too simplistic to describe what happens in the real world.

John Maynard Smith appreciated this. He took red deer as an example of where things go wrong for sexual selection theory. The powerful males get busy rutting in an exhausting, drawn-out, and impressive display of antler bashing. Often, though, the females aren't impressed and slope off to have sex with the less macho males of the herd. In a stroke of typical genius, Maynard Smith labeled them the *sneaky fuckers*.

Is it even sneaky? Perhaps it just makes good evolutionary sense. There isn't strong evidence that females really are impressed by the antler bashing or link it with the good genes that they supposedly seek for their offspring. And are there really so few good genes out there that the females are willing to focus all their attention on just one or two males? After all, if the theory holds together, all the males are the progeny of strong, fit males from the previous generation. It's hard to imagine that there is such a marked difference that females would be so discerning. The issue, known as the *Lek Paradox*, is well known to biologists. Although there are some explanations for why female choosing should persist, it still stands as a point of contention in standard sexual selection theory.

There are more examples of problems with the standard theory. Two Australian researchers, Mark Blows and Rob Brooks, found that the kinds of

selection done by fruit flies, for example, goes almost in the opposite direction to the one that sexual selection theory would predict. And the same researchers' studies with guppies showed the females are often lazy, not
making an effort to choose their mate carefully, but just mating randomly.
Others choose, but apparently on the basis of past experiences rather than
genetic traits. There are those who put in some effort and scrutinize the
males, but it is by no means the norm. As the biologist Steven Rose pointed
out, although it seems like a compelling idea, the empirical evidence for sexual selection based on impressive male traits is weak—and that is even true
among peacocks, the "classic" example. What's more, there is evidence suggesting that the key to reproductive success lies somewhere other than a display of brute strength.

In the summer of 1994 Elisabet Forsgren spent a couple of months playing matchmaker at the Klubban Biological Station on the west coast of Swe-
den. She was studying sand gobies, fish that swim around the shallows of
European shores, that she had caught in a shallow sandy bay and put into
tanks at the station. The fish dined on fresh mussels that Forsgren provided;
in return, they showed her just how complicated sexual selection can be.

First, Forsgren let two males fight each other for the best egg-laying site.
The winner was usually the slightly larger fish. Then she gave them a brood
of eggs to guard from a marauding crab. The smaller fish turned out to be a
better guardian. Finally, she let a female choose between them. The female—
who knew none of what had gone on—nearly always went for the male that
was a better guardian rather than the larger, dominant male.

That's not to say there isn't some truth to the standard theory of sexual
selection. One much-cited example is the elephant seal: the males fight each
other for access to the females. The biggest, strongest male wins and gets to
mate. Over successive breeding cycles this has led to the male elephant seal
becoming much bigger and heavier than the female; since the biggest male
in a group gets to sire the next generation, that next generation's males are
going to be bigger than those of the last generation.

Nevertheless, in Roughgarden's view there are so many exceptions to this
idea that we should look elsewhere for an explanation of courtship displays.
Secondary sexual characteristics, such as the peacock's tail, might not be in-
dicators of good genes, but of general good health, she suggests. An animal

in good health will also be able to help raise and protect more offspring, and producing a larger number of offspring that grow to maturity also makes a contribution to offsetting the cost of sexual reproduction. This idea certainly fits with Forsgren's discovery that some female fish choose a better, not a bigger, male.

What's more, a failure to impress doesn't make the less desirable members of the group walk away; they just take on different roles. Animals not directly involved in reproduction are still often involved in the group's welfare and cohesion, gathering food, offering protection, grooming—perhaps in return for a chance to copulate later. Such bonding activities, Roughgarden suggests, might be the root of the homosexual behavior that is so ubiquitous in the natural world.

Bruce Bagemihl's ten-year labor of love, *Biological Exuberance: Animal Homosexuality and Natural Diversity*, reports that more than 450 species have been documented engaging in nonprocreative sexual behavior—including long-term pairings. Two male black swans, for example, have been observed setting up a nest together, hatching (stolen) eggs, and raising perfectly well-adjusted cygnets. Better than well-adjusted, in fact; homosexual swans have a higher success rate in raising young than do heterosexual pairs.

Roughgarden has supplemented Bagemihl's work: in her book *Evolution's Rainbow* she took the total number of vertebrate species observed in "nonstandard" couplings up to three hundred or so. Many more examples may yet be exposed. Bagemihl's work took a decade partly because biologists suppress reports of homosexual behaviors in the natural world. One biologist told him that admitting the animals he was observing were living in a homosexual society was "emotionally beyond [him]." Others admitted documenting homosexual behavior in animals but not publishing until they had tenure.

These couplings certainly do not fit with the mainstream idea that genes, or at least organisms, are hell-bent on reproducing themselves. They do fit, however, with the idea of a social role for sex, and they fit with the idea that sexual reproduction is a spandrel, a by-product of some other phenomenon.

If Roughgarden is on to something, she believes it could have cultural as well as scientific implications. The orthodoxy of biology has corroded our culture like battery acid, she says. In general, we play out the roles prescribed for us by that culture—aggressive male and coy female—because deviation from its "norm" results in emotional and physical violence, bigotry, personal guilt, and criminalized behaviors. If biology has been getting it wrong, though, the new orthodoxy could trigger an infusion of tolerance; perhaps the anomalous prevalence of sexual reproduction will end up having deeper repercussions outside of science than within it.

Not everyone is convinced by Roughgarden's argument, however—indeed, most aren't. "I find this no less—but no more—compelling a theory than sexual selection, at least for social species," Steven Rose wrote when reviewing *Evolution's Rainbow* in the *Guardian*. Nevertheless, at the moment, evolutionary theorists need to look at all comers in considerations of sexual reproduction, and social selection is an intriguing possibility.

Easily the most intriguing thing about this possibility is that, if death is the root of sex (sex being necessary for life in an oxygen-rich environment), and passing on genes to the next generation is a spandrel and not the primary driver in the natural world, then it may be that, in evolution, group selection is not the perversity that Dawkins pronounced it to be. That would bring Joshua Mitteldorf's take on death—that it evolved from its initial appearance as a feature of eukaryotic life into a system that makes room for new generations—back into the realm of the possible. Mitteldorf's view is essentially the same one that August Weismann came up with in 1889 (but subsequently disavowed), so it could be said that by changing our view of sex we might also clear up the dark matter of death—with the first and most obvious theory. It almost seems too easy, but perhaps the answer has been staring us in the face all along. Could it be that sex is not the most important thing in life, and that group selection lies behind both sex and death? Can we solve two anomalies in one?

IF the descent toward death, and the subsequent rise of sex, began in the oceans, the tale of the female octopus provides a fitting conclusion to the story—and a nod toward our next anomaly. This creature is George Williams's dream organism, a tentacled testament to the power of antagonistic pleiotropy. She spawns once in her life and then loses the will to live; within ten days of her brood's hatch, she has starved herself to death. And this is death by programming. In 1977 the psychologist Jerome Wodinsky removed a female octopus's optic glands after she had spawned, preventing the hormone secretion that precipitates the self-starvation. Having thus been deprogrammed, she went on to live a long postreproductive life.

The female octopus is—quite literally—a martyr to her hormones. But we are no different. If we think we choose to eat, or get out of bed in the morning; if we think we choose to do anything much at all, we are sorely mistaken. The illusion—rather, the delusion—of free will is our next anomaly. And, perhaps, our most disturbing.

11

FREE WILL

Your decisions are not your own

In the spring of 2007, in a basement laboratory in central London, I played Pinocchio to Patrick Haggard's Gepeto. Haggard, a professor at University College, London's Institute of Cognitive Neuroscience, held a contraption that looked like an enormous cartoon key, something you'd use to wind up a clockwork mouse the size of a human being, over the left side of my skull. When he got the position right, he pressed a foot pedal and my right index finger moved. He slid the key along a bit, and my middle and then third fingers twitched. If he had mapped my skull properly, and turned up the power, he could have moved my leg or my arm. With this key, he can do almost anything.

This trick is a favorite tool of neuroscientists. It is called *transcranial magnetic stimulation*, and it uses two electrical coils to create a magnetic field that induces currents in the brain. With it, researchers can investigate the functions of particular areas of the brain. Haggard does this on himself a lot, he says. I was happy to experience it just this once. I don't really like it when someone else has control over my body.

I should count myself lucky, though; some people have to live with this lack of control on a daily basis. Those who suffer from *alien hand syndrome*, for example, can find themselves fighting one hand with the other. One of their hands, they often report, has a "mind of its own." They might be trying to put a cup down with their left hand and find the right hand is trying to pick it up. Or they are buttoning a shirt with their left hand while the right hand undoes the buttons. In extreme cases the alien hand tries to strangle the person; only a fight with the other hand saves them. These unfortunates sleep with their alien hand tied to the bed. Just in case.

Peculiar as this is, it has a straightforward explanation. It arises from lesions in the patient's brain. There are plenty of other examples: the man whose brain tumor turned him into a pedophile; the man whose damaged brain meant he famously mistook his wife for a hat. The lesson we learn from all this is that our minds do not exist separately from the physical material of our bodies. Though it is a scary and entirely unwelcome observation, we are brain-machines. We do not have what we think of as free will.

This inference can be drawn from decades of entirely reproducible experiments, and yet it doesn't make sense. As human beings we are utterly convinced of our autonomy, our self-determination, our free will. Almost everyone you talk to will say that such experimental results are anomalous; they don't fit into the framework of our conscious experience. Talk to Patrick Haggard, though, and he will tell you the anomaly, the curiosity, lies in our self-deception, the illusion of free will that we cling to so tightly. Haggard is not alone; most neuroscientists agree with him. But a few are still clinging to free will and casting the experimental results as the anomaly. The stakes in this fight couldn't be higher. Something about free will certainly doesn't make sense, and the resolution of this anomaly will determine what it means to be a human.

TELL most people they don't have free will, and they will defiantly tell you you're wrong. "Man defends himself from being regarded as an impotent object in the course of the universe," Albert Einstein wrote in 1931. If his disciplines, astronomy and cosmology, are leading the way in pushing human beings away from the center of the universe, the other sciences are not far behind, and free will is just about all that is left to mark us humans out as special. Even this may soon be lost, however.

In 1788 the philosopher Immanuel Kant put the problem of free will on a par with God and immortality. These, he said, were the only three things beyond the power of human intellect. Kant may have been wrong, however; little by little, neuroscientists are learning how to pull aside the curtain.

The first person to tear a hole in the illusion of free will was Benjamin Libet. Libet, who died in 2007 at the age of ninety-one, is a legend in neuroscience. But not, perhaps, for the reason he would have liked.

In the late 1970s Libet was in a round-table discussion on free will with the Nobel Prize–winning physiologist John Eccles. Eccles referred to a recent finding that a brain signal that precedes any voluntary action, called the *readiness potential*, kicks off a second or more before the action. At the time, Eccles believed that conscious free will initiates any and every voluntary action. Therefore, he said, conscious will must precede a voluntary act by at least a second. Immediately Libet recognized that this was a statement of faith; there was no evidence to back it up. So he went in search of the evidence.

Libet took a group of volunteers, wired them up with some scalp and wrist electrodes, and asked them to perform a very simple task. They had to stare at a clock and flick their wrists whenever they felt like it. Then they were to report when it was that they were first aware of the intention to make the movement.

With the scalp electrodes, Libet measured the steadily climbing signal of the readiness potential. The wrist electrodes gave precise timing for the muscle activity. When the subjects gave their timings for awareness of their intention to move, the intention always came before the action.

So far so good. But that's as far as the good news goes. Libet found that the brain's preparatory work, the readiness potential, preceded conscious intention—and by a lot. The brain was getting ready for the movement up to half a second before it happened, and on average that was 350 milliseconds before the subject was even aware he was going to move. By the time the subject experienced a conscious intention to move, his brain was going full speed ahead. Whatever he thought he was consciously deciding to do, it wasn't to make that movement.

Libet was completely taken aback by this discovery and immediately sought to rescue human free will in the only get-out he could find. There is time in between awareness of the intention to act and the action itself, he said, for a veto. We can make a conscious decision to not follow through with the action our brain is about to perform. And thus the lines were drawn in the battle for the essential nature of humanity.

ON the wall of Haggard's office is a piece of verse written by his daughter. It is called "A poem for Dad" and describes the reasons why she loves him. To a child, a parent's love is taken for granted; the child, though, has feelings that he or she feels can be rationalized and justified. Haggard earned his daughter's love by doing things, she says in her poem: helping with her homework, taking her swimming, and so on. Most of all, though, she loves him because he loves her.

Is this how machines behave? Do we really want science to be allowed to reduce human behavior—swimming, homework, love—to the firings of neurons that are independent of any individual's conscious will? And then there is the issue of right and wrong; we have built our civilizations, religions, and societies on the concept that people ought to be held responsible for their actions. Surely we only want to develop a scientific theory of human will if it legitimizes our concepts of moral responsibility? That was certainly Libet's view—especially since, he felt, his experiment might have been flawed. "The intuitive feelings about the phenomenon of free will form a fundamental basis for views of our human nature," he said. "Great care should be taken not to believe allegedly scientific conclusions about them which actually depend upon hidden *ad hoc* assumptions." He suggested that any theory that denies free will is "less attractive" than one that accommodates it. Unless there is some further evidence to the contrary, why not simply "adopt the view that we do have free will"?

Libet was right on one count, at least. The idea of free will has certainly not been killed stone dead by neuroscience; the protocols behind Libet's experiment are too loose for that conclusion to be drawn. While we talked in his second-floor office, Patrick Haggard had put a laptop computer on the table in front of me. I should try a version of Libet's experimental routine, he said. That, more than anything else, would show me why Libet's experiment has not yet put a definitive end to free will.

There certainly are difficulties with the experiment. In Haggard's version I have to press the F9 key while using a fast-spinning digital stopwatch on the screen to mentally note the time I am "aware of the will" to move my finger. There is plenty of room for experimental error here. How, for instance, do I get over my desire to press the key when the clock reaches a certain point in its cycle? And how do I disentangle my perception of the clock reading when I decide to press the key from my perception of its reading when I feel my finger press it? What does it even mean; how do I define "aware of the will to move"?

Many people have been here before me, Haggard says. To counter the first problem, a researcher carrying out the experiment tells the subjects over and over again that *they* are in charge, not the clock. Then they test the data, looking for patterns in timings that might skew the results. The second objection is more interesting and involves something called *cross-modal synchronization.*

If you have ever watched a badly dubbed movie, you will have experienced an annoying difficulty in following the dialogue. That arises because of problems with your cross-modal synchronization. You are watching the actors' lips move, and your brain is taking in this visual input quite happily. The trouble is, the audio input comes in through a separate channel. Your brain knows that it is easiest to understand speech when you have the visual input—the lip-reading—so it attempts to put the two channels, or modes, together.

Your brain is surprisingly forgiving here. If the soundtrack is out of sync by around 50 milliseconds, it doesn't matter; your brain can't tell. That's the level of error you're allowed when dubbing a movie; anything more, and people will start throwing things at the screen.

The same is true when Libet's subjects are synchronizing their view of the clock with their awareness of intention. The awareness is an internal mode, while the clock reading comes through the visual mode. Tests show the errors people make in synchronization are between 50 and 150 milliseconds. And that is nowhere near big enough to close the 350-millisecond gap between the unconscious initiation and the conscious urge to perform a movement.

Haggard is convinced there is no such thing as free will. The third objection, defining "aware of the will to move," is problematic, Haggard admits. But, he says, we're arguing semantics now; I'm playing a fool's game to try to close the gap by disputing the details of the experiment. It's there, he says, get used to it. Yes, the experiment has lots of flaws. Yes, it's not the perfect way to pin down the exact nature of voluntary versus involuntary action. But—and he is on the offensive now—what is the alternative? Do I really think I have free will? Do I really think that conscious thought can make my brain do things? Where is this thing, somewhere within my physical brain, that would make my brain leap into action and move my finger? There's no escaping it, Haggard says: our conscious "intentions" are by-products of something that is already going on. Proving this beyond doubt is difficult, of course. But, in Haggard's mind, one man has come closer than any other. And it is not Benjamin Libet.

In the early 1990s Itzhak Fried, a neurosurgeon at Yale University School of Medicine, was operating on the brains of patients with severe epilepsy. Their condition was so bad that part of their brains was to be cut out in order to stop the debilitating rapid fire of the neurons. To find out which neurons to excise, Fried attached a grid of electrodes to certain regions of the brain's surface; the idea was to monitor the neurons for overactivity.

Besides its clinical use, the situation also provided an unprecedented opportunity to fire up small regions of the brain with an electrical current to see what happens. It was a mapping opportunity, if you like, something that could help advance our understanding of how the brain works. Fried grasped this opportunity with both hands—and gained some unexpected results.

Altogether, Fried and his team stimulated 299 brain sites in thirteen patients; 129 of those sites gave a response. Most of those responses were simply movements of the body. I say *simply*, as if that weren't extraordinary enough. Fried and his team were applying currents to specific regions of the brain and evoking movements—sometimes just one joint would flex or one muscle group in the face would contract. Sometimes they could evoke a larger response: the patient would adopt a certain posture, extending her neck then rotating her head to the right, for example. That is, by any standards, extraordinary.

But it wasn't the most extraordinary thing. What really shocked the researchers was the patients' reports that they were feeling "urges." An urge to move my right arm. An urge to move my right leg inward. An urge to move my right thumb and index finger. And when the researchers ramped up the current a little on each case, that's exactly what happened: the urge turned into the action, the very action the patients had reported wanting to perform.

All this at the flick of a switch. The researchers had taken over the patients' will, and then—by giving it a bit more juice—they took over their body.

I could tell, as he described them, that Patrick Haggard is enthralled by these findings. "It would be riveting to have this done to you," he says.

He doesn't want anyone tinkering directly with his brain, though—which is why we ended up in his basement lab. Transcranial magnetic stimulation is an indirect, and consequently less effective, version of what Itzhak Fried did to his epilepsy patients. But, in essence, it is the same.

I have to admit, watching Haggard move my finger strained my sense of self to the limit. That digit seemed to me like somebody else's finger. Nevertheless, it was instructive: it showed me something more about the Libet experiment. Whatever my problems over the phrase "aware of the will to move," there is a big difference between a movement that comes from your own conscious intention and a movement that comes from—well, seemingly, nowhere at all. It's not a reflex, like ducking a low-flying pigeon in Central Park or straightening your leg after a doctor taps you below the kneecap. It's not like hitting a fast-moving baseball. All those things feel like human capabilities; I might not know how I do them, but at least I know it is me doing them. This was different. It wasn't me. Being Patrick Haggard's puppet was quite a revelation; I became ever more convinced that I don't have free will.

The neuroscience literature attacks the free will delusion from another angle too: neuroscientists have shown time and again that when it comes to intention and control, we are astonishingly self-deceiving. We might be convinced that we have free will, but we should treat any and all such inner convictions with a large dose of skepticism.

Daniel Wegner and Thalia Wheatley proved this in 1999 with a customized version of what they rather entertainingly called an "ordinary household Ouija board." The pair were based at the University of Virginia at the time and decided to test psychology students' beliefs about their control of their hand movements. The students gained a course credit for taking part; the researchers gained a much-cited classic result.

The experiment involved deception from the start. Each student arrived for the experiment at the same time as someone who was in on the trick. The student believed this insider was also a naive participant, and proceeded to work alongside that individual.

The Ouija board was a computer mouse with a square piece of board glued on top; the pair were to place their fingertips on the side of the board closest to them. They were then instructed to move the mouse together, in slow sweeping circles that moved a cursor around a computer screen. The screen showed fifty small toy objects: a swan, a car, a dinosaur, and so on. Every thirty seconds, they were to stop moving the mouse and individually rate how much it was *their* intention to make it stop there.

The scam was complex, involving covert instructions to the insider, but the result was clear. Though all the cursor movement and all the stops were due to the insider, the students reported that the stops were their intention. They believed themselves to be making the decisions when it was clear to everyone else that they weren't.

Wegner also carried out related experiments that asked students to "read the unconscious muscle movements" of their student partner. In these studies, the students were under the impression that they and their partners both heard simple questions such as "Is Washington, D.C., the capital of the United States?" The students had their fingers on top of their partners' fingers and had to "feel" their partner's response, then press the appropriate key: yes or no.

In reality, the partner—an insider again—heard nothing and thus made no response. The students got the answers right 87 percent of the time—but attributed the answers to the influence of their partner 37 percent of the time. In other words, the correct answers were often produced automatically, without conscious contribution. An expectation of their partners' unconscious movement was enough to undermine the experience of conscious will.

The conclusion? Our perceptions, actions, and intentions are dangerously malleable. We are like small children sitting in front of an arcade race game; even if no money has been put in, and the cars on the screen are racing in demo mode, they grab the steering wheel, move it back and forth, and believe they're driving. Wegner and Wheatley think these kinds of phenomena lie behind the skills of many stage entertainers. "Believing that our conscious thoughts cause our actions is an error based on the illusory experience of will—much like believing that a rabbit has indeed popped out of an empty hat," they wrote in the July 1999 issue of *American Psychologist*.

It is likely that shows involving hypnosis, mind-reading, and illusion all utilize our shaky grip on the real nature of conscious free will. Set things up right, and you can trick people into thinking they are causing something to happen. Alter the setup, and you can trick people into thinking someone else is controlling their behavior. Or that they have carefully watched every part of a sequence of events. Theaters across the world provide the laboratories that prove this idea: under the supervision of showmen and illusionists, thousands of people have moved a glass around a Ouija board with no awareness that they are doing it themselves. Proof of just how extraordinarily resistant to reality we human beings are comes with the knowledge that for almost the whole time that illusionists and fraudsters have been profiting from this phenomenon—a century and a half now—we have had a perfectly good, rational, and spirit-free explanation for it: *ideomotor movements*. These are tiny unconscious motor movements that arise and are amplified through concentrated expectation of movement. They were first identified as the "influence of suggestion in modifying and directing muscular movement, independently of volition" in 1852 by the psychologist William Benjamin Carpenter. The result is large movements that the subject has no awareness of causing.

The psychologist and philosopher William James, brother of novelist Henry, took Carpenter's baton and ran with it, carrying out experiments to show just how easy it is for us to bypass our volition. In 1890 he laid out his findings in *The Principles of Psychology*, where he stated that "every mental representation of a movement awakens to some degree the actual movement which is its object." If there is nothing to stop it, he said, the movement grows.

James was the first to realize that not all of our delusions of control are quite as otherworldly as the Ouija board effect. He pointed out that something as simple as getting out of bed in the morning can be similarly problematic. In fact, James considered the action of getting out of bed to "contain in miniature form the data for an entire psychology of volition." Perhaps it takes a rather unconventional mind to see getting up as so laden with meaning. James was certainly unconventional; he used drugs such as amyl nitrate and peyote in his study of mystical experience (and contended that only under the influence of laughing gas did he ever understand the philosophies of Hegel). His observation of how hard it is to get up in the morning, however, is rather insightful.

> We know what it is to get out of bed on a freezing morning in a room without a fire, and how the very vital principle within us protests against the ordeal . . . now how do we *ever* get up under such circumstances? If I may generalize from my own experience, we more often than not get up without any struggle or decision at all. We suddenly find that we *have* got up.

It is a startlingly obvious, yet almost universally ignored, example of a lack of conscious control over actions. We've all had the experience: it's 7:15 a.m.; rise-and-shine time. You're lying under the duvet, listening to some radio announcer telling you that it's a beautiful day out there and the traffic over the harbor bridge is running smoothly. There's no reason to stay in bed. You tell yourself to get up. It doesn't happen. Then, miraculously, thirty seconds later, you find you have done it. You don't remember reissuing the command, but there you are, standing by the window, gazing bleary-eyed out into the sunshine. You routinely operate without conscious control.

THE idea of free will goes to the center of our sense of self, our autonomy as human beings. Strip us of it, and we are nothing more than animals. This, perhaps, is what is most disturbing about the fate of Alex, the narrator in Anthony Burgess's novel *A Clockwork Orange*. For all the "ultraviolence," for all the rape and theft and bloody beatings he doles out, it is his punishment that is most unsettling. Alex undergoes conditioning, reprogramming, so that he responds to violence with unbearable nausea. He winds up unable to perform the sadistic acts he enjoys; he no longer has the choice of whether to do good or evil. The prison chaplain has deep misgivings about the process. "When a man cannot choose, he ceases to be a man," he says. "Does God want woodness or the choice of goodness?"

Writing in *American Scientist* with Sukhvinder Obhi, Haggard put it another way: questioning our free will risks a "philosophical firestorm." Haggard knows, however, that the philosophical firestorm will be nothing compared to the legal firestorm that is coming.

Brain scanning is becoming extremely sophisticated. It is no longer about finding which area processes vision or which area controls the motor functions. Neuroscientists are now identifying the seats of attributes we associate with the person, not the organism. Guilt, shame, regret, loss, impulsivity—they are all measurable entities. The anatomy of personality and experience is being reduced to electrical signals. If we find some people are hard-wired for impulsive behavior—and we are beginning to get there— how long before it is cited as a legal defense? How long before neuroscientists testify that someone cannot be held responsible for the way his brain circuits are connected? Haggard has yet to testify in court. He has been asked, but he has never felt he could offer a "clear, valid, and useful" contribution to the case. No one wants to wander casually through this territory, it seems.

David Hodgson certainly doesn't. Hodgson, a legal philosopher based in Sydney, Australia, argues, like Libet, that free will is too essential a part of humanity to let our limited scientific understanding remove it at this stage in the endeavor. Hodgson thinks that, though we have some evidence to the contrary at the moment, future experiments may well confirm our free will. Henry Stapp, a physicist based at the Lawrence Berkeley National Laboratory in California, cites quantum theory as a source of doubt on the experimental evidence of the Libet experiment. In quantum theory, the act of observation can change the experimental conditions, so the results of any experiment that involves self-observation cannot be taken at face value.

Such skeptical viewpoints are certainly in the scientific minority. They are based on the scientifically indefensible premise that we simply *must* have free will, and that any experimental results that show otherwise must be flawed. On the other side of the fence, the British psychologist Guy Claxton thinks clinging to free will is akin to denying that the Earth goes around the Sun. Yes, a heliocentric universe is somehow a less comforting worldview; yes, it makes us feel less special. What's more, yes, you can live quite happily without it, as people did for millennia. The only time it really doesn't work is when you want to do something complex, like leave the planet.

Similarly, Claxton says, it is only OK to believe you have free will if you don't try to do anything complex like control everything in your life. Studies show that neurotic and psychiatric disorders are more common among those who attempt to keep conscious control of life and suppress its unwelcome quirks. Sanity, paradoxically, may lie in accepting that you are not in control.

It's easier said than done. We are ill-equipped to live ultrarational lives; psychologists have repeatedly shown that our ideas of "rational" decision making are often self-delusion. In one of the most-cited papers in psychology, for example, Richard Nisbett and Timothy Wilson showed that we are unable to explain even why we choose to buy one particular pair of socks over another. Wilson also showed that decisions we think long and hard about are the ones we end up less happy about. So it's likely that thinking long and hard about free will and making a "rational" decision about it based on the evidence is not even a great idea. If you've got this far in the chapter, you're probably not going to be happy whichever side of the fence you come down on. It might be best to continue with the wishful thinking you began with; the best advice, after all these arguments and demonstrations, must surely be: do nothing. Free will may be the one scientific anomaly that humans would be wise to ignore.

For all practical purposes, it makes sense to retain the illusion. Human consciousness, our sense of self and intention, may be nothing more than a by-product of being the enormously complex machines that are our big-brained bodies, but it is a useful one, enabling us to deal with a complex environment. What's more, our human cultural arrangements have evolved in parallel with our consciousness, and they rely on the naive view that we are able to direct (and are thus responsible for) our own actions. Philosophers will continue to discuss the implications of the scientific facts with sangfroid, but coldly conceding we are brain-machines and giving up on the notion of personal responsibility will most likely remain too dangerous a move for those having to deal with real-world situations. There is surely too much at stake—too many unforeseeable consequences—to risk dismantling our societal norms for the sake of scientific "truth." Taking the ultrarational option might get us nowhere—and that would most likely be the best result we could hope for. More likely, the destruction of our legal and cultural frameworks in the light of scientific revelations would take us somewhere we really don't want to go. It is possible that if invoked in legislation, our scientific efforts could undermine some of the foundations on which human society has been constructed. The Harvard University psychologist Steven Pinker probably put it best. "Free will is a fictional construction," he said. "But it has applications in the real world."

IN the illusion of free will, it seems we have been equipped with a neurological sleight of hand that, while contrarational, helps us deal with a complex social and physical environment. This is not the only mind trick that evolution has bestowed upon us. There is another neurological anomaly that sits beyond our conscious control, and it is certainly too late to leave this one alone; it has already been scientifically deconstructed and set as a central pillar of our health-care system, arbitrating what works and what doesn't in modern medicine. It is the placebo effect.

12

THE PLACEBO EFFECT

Who's being deceived?

"It has brought me great comfort to know that I could, in some way, help people feel better," said Leo Sternbach, inventor of the antianxiety drug diazepam. Sternbach certainly did that—in spades. What is only just starting to emerge is just how much Leo Sternbach's drug depends on people helping themselves to feel better.

From 1969 to 1982, diazepam, marketed as Valium, was the top-selling pharmaceutical in the United States. At the height of its powers, Sternbach's employer, the pharmaceutical giant Hoffman LaRoche, sold 2.3 billion of the little yellow pills marked with a *V*. That was in 1978, and the drug had already been part of popular culture for more than a decade; "Mother's Little Helper" by the Rolling Stones, released in 1966, is a satire on domestic abuse of Valium. In the same year that song was released, the drug gained a starring role in the cult novel *Valley of the Dolls*; diazepam "dolls" were the lead characters' means of getting through the strains of life in New York. Diazepam is now, according to the World Health Organization, a "core medicine," essential for any nation's pharmaceutical store. The strange thing is, it doesn't work unless you know you're taking it.

In 2003 a paper in *Prevention and Treatment* reported that diazepam had no effect on anxiety when it was administered without the patient's knowledge. In an extraordinary experiment, researchers in Turin split a group of trial subjects into two. One half were given diazepam by a doctor who told them they were being given a powerful antianxiety drug. The other group were hooked up to an automatic infusion machine and given the same dose of diazepam—but with no one in the room and no way of telling they had received the drug. Two hours later, the people in the first group reported a significant reduction in their levels of anxiety. The second group reported no change. "Anxiety reduction after the open diazepam administration was a placebo effect," the researchers suggested.

A placebo is a medical procedure that has no medicine in it. A sugar pill, or a spoonful of sugar water, a saline drip—or anything, really. A parade of doctors in white coats coming to your bedside to offer reassurances can be enough to trigger the effect. The power of placebo comes from the deceptive message that comes with it. You are told (or you sense) this procedure or ritual will have an effect on your body or state of mind, and if you genuinely believe it, taking the pill or the drink, or in some cases just seeing the doctor, will produce exactly that effect. Witch doctors, shamans, and other purveyors of the magical arts are known to deal in placebos. When they carry out a sham ritual to cure a paying believer, that cure can work wonders. The same might be said of televangelists. And Western medical doctors, too; research has shown that white coats and stethoscopes can produce surprisingly effective placebo effects—as can a good bedside manner. Doctors know that if patients feel they are getting a suitable treatment, the treatment is enormously more effective.

In one sense there's an easy explanation for all this: the chemistry of the drug is being augmented by chemicals secreted in the brain—the effect of what Fabrizio Benedetti, the leader of the Turin group, calls "the molecules of hope." The difficult side of the new experimental evidence is that, where we once thought we had a handle on the placebo effect, it is now becoming clear that we don't.

In medicine, we have long been accustomed to accounting for placebo. Modern scientific medicine was constructed on the notion of the *randomized double-blind, placebo-controlled trial*, where drugs have to perform better than a dummy pill or inert saline injection. Now, though, things aren't so clear. Some analyses of the data suggest that the placebo effect is largely a myth. What's more, the medical system was set up assuming not only the existence of placebo but also that its effects can be separated out from the chemistry of the drugs being tested. It seems that assumption was false, and the edifice of the pharmaceutical trial may have to be dismantled. No wonder a recent National Institutes of Health conference declared placebo research an "urgent priority."

Benjamin Franklin, the father of rational, "evidence-based" medicine, must be turning in his grave. In 1785 Franklin headed a commission to investigate the claims of "animal magnetism." The Austrian physician Franz Anton Mesmer had entranced (hence *mesmerized*) Paris with his claims that magnets and glasses of water could be used to healing effect. Louis XVI wanted to know whether these claims stood up, and some of the greatest scientists in Europe were commissioned to find out the truth. Their tests were the first scientific inquiries to use blindfolds that prevented the subjects from biasing the results—the original "blinded" trials really were just that. The commission's report came out in 1785. Any healing effect is "really due to the power of the imagination," it said.

Interestingly, 1785 was also the year the term *placebo* appeared for the first time in a medical dictionary. It was the expanded second edition of George Motherby's *New Medical Dictionary*, and the word, to Motherby, meant "a common place method or medicine." Though that is not particularly damning at first glance, it was most likely a negative label, meaning the medicine was trivial, or unimpressive, because the word already had a negative connotation. *Placebo*, which means "I will please," had come to signify insincerity, flattery, and profiteering since medieval times, when greedy churchmen would take mourners' money to sing Psalm 116 at funerals. The psalm begins, *Placebo Domino in regione vivorum* (I will please the Lord in the land of the living). By 1811, that negative connotation was well established; Robert Hooper published his *New Medical Dictionary* with an entry for *placebo* that read: "an epithet given to any medicine adapted more to please than benefit the patient." Little did the clinicians of Hooper's day know that a placebo might benefit patients just as much as it pleased them.

As often happens, that knowledge had been gained and lost before. It was certainly known to the ancient Greeks. In 380 BCE Plato wrote *Charmides*, in which the Thracian king Zamolxis tells Socrates that the great error of the physicians of his day was the separation of the soul from the body. Despite doctors' best efforts, curing the body is impossible without flattering the mind, Zamolxis says.

> If the head and body are to be well, you must begin by curing the soul;
> that is the first thing. And the cure, my dear youth, has to be effected by
> the use of certain charms, and these charms are fair words; and by them
> temperance is implanted in the soul, and where temperance is, there
> health is speedily imparted, not only to the head, but to the whole body.

Plato was right; words are powerful. If you communicate that you are doing something—if you utter what the French psychiatrist Patrick Lemoine calls the *incantation*—it can work wonders.

An example of an incantation, drawn from Lemoine's experience, might be, "I'm going to prescribe you some magnesium that will treat your anxiety." Magnesium isn't a licensed cure for anxiety, but magnesium deficiency produces symptoms similar to anxiety; in a bizarre nod to the principles of vaccination, European clinicians often prescribe magnesium for anxiety, Lemoine says. And not only are his patients satisfied; they get better—and relapse if the treatment is interrupted. Nearly 250 years into the era of evidence-based medicine, the incantation is still a powerful force.

A 1954 paper in the *Lancet* declared that the placebo effect is only useful in treating "some unintelligent or inadequate patients"; that seems almost laughable now. According to Ann Helm of the Oregon Health Sciences University, somewhere between 35 and 45 percent of all medical prescriptions are placebos. That estimate was made in 1985. In 2003 a survey of nearly eight hundred Danish clinicians, published in *Evaluation and the Health Professions*, found that almost half prescribed a placebo ten or more times per year. A 2004 study of Israeli doctors, published in the *British Medical Journal*, determined that 60 percent had prescribed placebos, more than half of them doing it once a month or more. Of the Israeli doctors who prescribed placebos, 94 percent said they found them to be an effective means of treatment.

These are not pure placebos. The doctor can't send you to a pharmacy to get a sugar pill; after all, you might read the prescription, breaking the spell. No, doctors routinely prescribe medications that have a tiny bit of something useful in them—but its licensed use is not to treat what is ailing you.

Despite being so commonplace, it is a practice that splits the medical community. It is seen by some as unethical—dangerous, even. And not only is it practicing deception on a patient; it also forces other medical professionals to act as accomplices to the placebo-prescribing physician. After all, what do you do with your prescription? You take it along to the pharmacist. Your pharmacist then—willingly or reluctantly—tends to play along. An article in the *Journal of the American Pharmaceutical Association* even provides a script for their role. Realizing that a doctor has prescribed a placebo, the pharmacist should deliver the medication with these words: "Generally, a larger dose is used for most patients, but your doctor believes that you'll benefit from this dose." The pharmacist might then advise you of some possible side effects. Or not.

If this shocks you, you can be comforted by the fact that no one is out to fleece you. Neither your doctor nor your pharmacist is getting away with some scam. They are simply doing what they can for your health. They know that you have faith in their abilities; otherwise you wouldn't have come for the consultation. And their abilities include the knowledge that placebos work—though no one knows exactly why. You have faith in your doctor, and that faith can make you well. The nature of placebo simply means that they have to practice a tiny little deception to help it happen. Is that wrong? There is no consensus on the answer to that question.

WHILE the ethical issues surrounding placebo have long been debated to no conclusion, the scientific basis of the effect is a relatively new topic for research. The general conclusion here, it seems, is that the placebo effect is due to chemistry. The classic demonstration involves inducing pain in subjects; the original work was done by dentists who had extracted molars from patients. However, less drastic measures are possible. The only truly essential ingredient is a little deceit.

It all kicks off with the pain-racked patients receiving something like a morphine drip. Later, after the patients have begun to associate the morphine with pain relief, you can subtly substitute saline solution for the morphine. The patients don't know their "morphine" is nothing but salt water and, thanks to the placebo effect, they report that their pain medication is still working fine. That is strange in itself, but not as strange as the next twist makes things. Without saying anything to the patients, you put another drug into the drip: naloxone, which blocks the action of morphine. Even though there is no morphine going into the patients' bodies, naloxone still stops the pain relief in its tracks; the patients, oblivious to all that has gone on, now report that they are in discomfort again.

The only plausible explanation is that the drug that blocks morphine's pain-relieving power also blocks the saline's (placebo-based) pain-relieving power. Which means the saline really was doing something—it wasn't all in the patient's imagination. Or at least it means that imagination can have a physiological effect.

When the dentists first performed this trick, they attributed the placebo effect to a stimulation of the body's endorphins, natural opioids that act using the same biochemical pathways as morphine. The expectation of pain relief was enough to trigger an endorphin release that did the job, they concluded. Then the naloxone blocked the endorphins; that's why the pain came back. It turns out to be more complicated that that, however.

What was once considered nothing more than the fancies of the imagination is a real, repeatable, and multifaceted biochemical phenomenon. The placebo effect pulls out all the stops; the expectation of pain relief can stimulate all kinds of natural pain-relieving chemicals. Use ketorolac, a painkiller that works via a completely different chemistry from that of morphine, in the conditioning, then replace it with saline. The addition of naloxone does nothing there because the placebo pain relief is provided not by endorphins but by some other natural painkiller that your body produces. The stimulation of hormones that work in the same way as the painkiller sumatriptan is one example. The phenomenon even depends on how much pain the patient is expecting to feel. Tell ready-conditioned patients they are getting morphine that is more dilute than usual (when in fact they were getting nothing more than saline), and introduce naloxone. Again, it doesn't block the painkilling effect of the saline because the expectation of reduced pain relief has triggered some alternate mechanism. What everyone thinks of as "the placebo effect" turns out to be a whole array of different effects, each with a unique biochemical mechanism. Our brains can fool us in any number of ways.

THOUGH all this seems completely convincing—by now, we are surely confident that the placebo effect is a real phenomenon—there is a fly in the ointment. In 2001 two Danish researchers published a landmark paper in the *New England Journal of Medicine*. Asbjorn Hróbjartsson and Peter Gøtzsche had begun to get suspicious about claims of the efficacy of the placebo effect. Everywhere they looked—in textbooks, journal papers, and magazine articles—authors were quoting a number the pair couldn't quite believe. According to almost everything in the medical literature, 35 percent of patients would get better if told a dummy treatment they had been given was real.

Eventually, they found the source of this much-quoted, never-questioned statistic: Henry Knowles Beecher. In *The Powerful Placebo*, published in the *Journal of the American Medical Association* in 1955, Beecher made the first loud call for the use of double-blind, placebo-controlled trials in assessing medical treatments. The paper documents his analysis of a dozen studies, an analysis that produced the magical 35 percent figure.

It wasn't enough to convince Hróbjartsson and Gøtzsche, so they carried out a meta-analysis. This is what scientists do when they are faced with a long series of conflicting answers to a question; essentially, it is a formalized way of analyzing all previous attempts to answer the question. They examine the quality of each one: its experimental methods, its biases, its statistical analyses. The idea is to get a flavor of each set of results and then put them together in a way that reflects how much weight should be given to their stated results. In the end, such a study makes some pronouncement about the overall weight of evidence for or against a hypothesis.

THE PLACEBO EFFECT
171

Hróbjartsson and Gøtzsche's meta-analysis of the placebo effect took the data from 114 clinical trials that had compared placebo-treated patients with untreated patients. Overall, there were around 7,500 patients suffering from about forty different conditions ranging from alcohol dependence to Parkinson's disease. Over this wide spectrum of complaints, they found no evidence that placebo treatments had significant effects on health. The only place there was possibly some effect was in the trials that involved pain relief, but even here it was hard to be sure. Pain is a subjective measure, and patients like to please their doctors, Hróbjartsson points out; they may well have reported less pain than they actually felt. Certainly the objective measures, such as blood pressure and cholesterol levels, showed no placebo response. The researchers called for doctors to stop using placebos in clinical situations. "The use of placebo outside the aegis of a controlled, properly designed clinical trial cannot be recommended," they said.

In 2003 Hróbjartsson and Gøtzsche revisited the analysis, this time with data from 156 trials and 11,737 patients. Their results, published in the *Journal of Internal Medicine*, were unchanged. They "found no evidence that placebo interventions in general have large clinical effects, and no reliable evidence that they have clinically useful effects." Placebo, they conclude, is a far from proven phenomenon; the only possible exception is in pain relief, and even here the placebo response was not clearly above what they would expect to see in doctor-pleasing-biased subjective reporting. "Most patients are polite and prone to please the investigators by reporting improvement, even when no improvement was felt . . . we suspect reporting bias occurred," the researchers write.

Hróbjartsson and Gøtzsche's work is well respected and has contributed a significant amount to the controversy over our handling of placebo. Nevertheless, we have significant evidence from equally well-respected researchers that the placebo effect is real. Brain imaging has shown the pathways involved in the brain, for example. In 2005 researchers from the University of Michigan published their work with a *positron emission tomography (PET)* scanner, showing the endorphin system in the hypothalamus activating when patients received an injection they had been told was a pain medication. Reporting bias seems unlikely given that these trial patients were being deliberately hurt (by a saline injection in the jaw) as part of the experiment; they had no reason to report less pain in order to please the researchers carrying out the experiment.

An editorial accompanying Hróbjartsson and Gøtzsche's original paper seems to sum up the general feeling. Though the author, John Bailar of the University of Chicago, admitted to little more justification than a "pesky, utterly unscientific feeling that some things just ought to be true," he suggested their conclusions were "too sweeping." Things that happen in research labs "may obscure a real effect of placebo that would be evident in nonresearch settings." The solution to this problem is unforthcoming, however; "it is not clear how one could study and compare the effects of placebo in research and nonresearch settings, since that would of course require a research study."

Perhaps an informal visit to Turin would help Bailar. It certainly cured me of any doubt about the reality of the placebo effect.

WHEN I asked Fabrizio Benedetti if I could experience a placebo response for myself, he was far from convinced it would work. Normally, his team won't tell their trial volunteers what kind of experiment they are carrying out; such knowledge might skew the results. It didn't in my case. In a windowless basement room below Turin's towering San Giovanni Battista hospital, I repeatedly subjected myself to pain. And, against all my expectations and with my full knowledge of what they were doing, the doctors present were able to reduce it with nothing more than a lie.

My first experiment measured the effects of caffeine on muscle performance, following a routine that involves exercising before and after a small cup of cold, rather unpleasant coffee. While I was drinking the coffee, the white-coated Dr. Antonella Pollo, one of Benedetti's colleagues, filled my head with stories about how caffeine is a banned substance in athletics. Her sister, she said, does archery. She is always told not to drink anything containing caffeine before an event; apparently; it enhances muscle performance and gives an unfair advantage. I knew there was a lie somewhere— perhaps there was no caffeine in the coffee, maybe caffeine has no effect on muscle performance, or maybe Pollo simply reduced the resistant weights for the exercise session after the coffee break—but I was definitely able to do more after the coffee than before.

When the experiment came to an end, Pollo came clean. There was no caffeine in the coffee. Nevertheless, I had been sufficiently convinced of my increased powers to perform much better the second time around. She looked rather pleased. The experiment was far from rigorous and—in my quick and dirty clinical trial, at least—had many flaws. What's more, she hadn't expected it to work at all on someone who knew what was going on.

The next test came from another white-coated doctor, Luana Colloca, who entered the room holding what looked like a couple of button cell batteries on a plastic strip. They were electrodes. "Do you mind receiving an electric shock?" she said.

When I consented, she strapped the electrodes to my forearm. Then she wired me up to a computer programmed to manipulate the mind as it gives a series of electric shocks.

The computer screen told me—via a red or a green light—whether the shock I was about to get would be mild or severe. The deception here comes from a conditioning, where the brain learns to associate a color with an anticipation of a particular level of pain. The screen shows a color, and about five seconds later the computer gives a shock. Green for severe (something like an electric fence jolt) and red for mild (no more painful than a light touch on the arm). But once the conditioning is established, playing with the color can play with the brain's perception of pain.

It worked. After around fifteen minutes of conditioning, the last run of shocks all felt mild, like a touch on the arm, whether introduced under a red or a green light. But they were all severe, Colloca told me afterward. By rights, every one of them should have felt like touching an electric fence.

In some ways, I shouldn't be surprised. The brain is an astonishing organ, a supremely complex collection of molecules that process signals—both chemical and electrical—to give us our sense of who we are and how we experience the world around us. With careful control of the signals going in, why shouldn't that sense be open to manipulation?

We know there are many ways to alter the state of a human brain and the body it oversees. The most obvious are the five senses: the smell of cut grass evokes a particular memory state; the taste of chocolate releases serotonin;

the touch of a lover and the sight of a big-eyed puppy both release the oxytocin molecules that bond us to our partners or our children (or our dogs); the sound of a scream sends a rush of adrenaline through us, making us ready for fight or flight.

Electrical signals can bypass conscious bodily control too. Sufferers of Parkinson's disease, for example, can have their tremors stopped with a microchip implanted in the hypothalamus. Benedetti, an experienced neurosurgeon, performs such implantations; not only can they help a Parkinson's patient's motor control, but they also provide a tool for investigating the neural mechanisms of the placebo effect. Tell patients that their implant's settings have been altered so that it will be harder to control their movements, and they respond by doing everything at a snail's pace. Tell them the opposite—that the electrodes are now set for optimum mobility—and suddenly the movements become normal. In neither case does anyone need to touch the electrode settings to achieve the effect: expectation of a significant improvement—or degradation—in the motor control of Parkinson's patients gives them just that; tell them they're going to be impaired, and they will be. It's not just about positive thinking: it's about the chemical or electrical signals that positive thinking produce.

Benedetti has shown this explicitly. The classic Parkinsonian symptoms of muscle stiffness and tremors are caused by explosive bursts of signals coming out of a specific region of the brain: the subthalamic nucleus. Injections of the drug apomorphine reduce this hyperactivity to near-normal levels and take away the associated stiffness and tremors. Benedetti's team took a group of sufferers who had had electrodes implanted in the subthalamic nucleus, and gave them apomorphine injections for a few days. They then covertly switched the injection to saline—still telling the patients that the injection would relieve their symptoms. It did, and measurements through the implanted electrodes showed reduced activity in the neurons of the subthalamic nucleus. Placebo, it seems, is all in the brain—and it is real.

IT is here that the placebo effect turns into something like medicine's equivalent of dark energy: a repeatable, measurable phenomenon that could still turn out to be an illusion. A broad analysis of the best clinical data says it might not exist—at least not in significant amounts. But even with full knowledge of what was going on, I found myself powerless to resist the placebo effect. It is not simply about deception, a sugar pill being perceived as an efficacious cure. We can create it with mind tricks, brain implants, or chemical cocktails, and we can see it working on brain scans. Though there is scientific evidence that the placebo effect is a myth, or that we have misled ourselves about what is going on, there is perhaps more evidence pointing the other way.

Clinical studies show you can cut morphine use by half—over the long term—if you just make sure the patient knows you're giving it. Telling patients they are being injected with a painkiller—while injecting them with saline—is as effective as injecting 6–8 mg of morphine. Studies at the U.S. National Institutes of Health found that cocaine abusers in a recovery clinic can get by on half doses too—as long as they know they're getting something. Expectation is a powerful thing.

In fact, we're back at diazepam. On its own—administered covertly—it does nothing. It's about diazepam *plus* the expectation chemicals that anticipation of a dose produces; the expectation chemicals are quite good by themselves, but with diazepam added to the mix, you're really in for a treat.

These expectation chemicals have a dark side too, though. Benedetti and Colloca have already started to put warnings out that placebo research could be exploited for questionable purposes. We are only wading in the shallows of the science of placebo, and it's already clear that this, like genetics, could be a murky pond. "There are . . . potentially negative outcomes of placebo research," they wrote in a *Nature Reviews* article in 2005. "If future research leads to a full understanding of the mechanisms of suggestibility of the human mind, an ethical debate will then be required."

That is especially true in light of the *nocebo effect*, where deliberately inducing anxiety can make pain worse. Benedetti is one of the few people who have been able to study this phenomenon; if researching placebo poses an ethical dilemma for doctors, nocebo doubles it.

Nocebo means "I shall harm." In a nocebo study, the harmless medicine is delivered with a phrase such as, "This really will make you feel much worse." It could prove an extremely valuable tool, and Benedetti is already using his nocebo experience to overcome the limitations of current painkillers, but what kind of ethics committee gives approval to a scheme designed to make patients more uncomfortable through lying to them? None. Which is why Benedetti has to rely on paid volunteers who are willing to suffer.

It started in 1997, when he and his colleagues were testing the idea that anxiety makes pain worse. They injected a group of patients who were recovering from painful surgery with proglumide, a chemical that blocks the action of cholecystokinin (CCK), a neurotransmitter chemical associated with anxiety. When they gave these patients an inert pill and told them it would make them feel worse, it simply didn't. It was impossible to induce the nocebo effect when CCK was blocked.

It was a good result, but scientifically lacking—there was no control group that *didn't* get the CCK-blocking proglumide and thus *did* feel the additional discomfort that anxiety can bring. Unfortunately (for Benedetti, if not for the patients), there was no ethical approval for a control group.

It took Benedetti nearly ten years to get approval and volunteers for a follow-up study. At the end of 2006 his team published a paper showing that we—or rather our neurotransmitters—can turn anxiety into pain. The volunteers underwent a routine involving a tourniquet, some injections, and a verbal warning that their pain would increase while Benedetti's team took blood samples and asked them how they rated their pain. The blood samples gave the researchers what they were looking for: proof that proglumide stops us from turning chemical signals of anxiety into exaggerated pain. Proglumide is the only CCK blocker licensed for human use, but it is not particularly effective. When researchers manage to develop something better, they will have a drug that can be mixed with narcotics to alleviate physiological and psychological pain simultaneously. Though nocebo seems somewhat dark—one can imagine it being exploited to produce extra anxiety and thus pain in interrogations, for example—at least it has positive applications too.

FOR medicine, the placebo effect is a two-edged sword. Despite Hróbjartsson and Gøtzsche's results, it seems undeniably useful—but it also takes away our certainties. We can't tell what the chemical constituents of a drug actually do to the biochemistry of our bodies, because even the sight of the approaching needle starts to disturb the biochemical environment. It is, Benedetti says, like the uncertainty principle of physics: anytime you measure something, you necessarily disturb it, so you can't ever be sure that your measurement is accurate. As a result, it seems that we may have to redesign drug trials.

Our slowly unfolding understanding of the placebo effect means we may need to reinterpret all our pharmaceutical data. In some cases, clinical trial results will seem invalid, or will at least need to be taken with a pinch of salt. It has taken decades to refine our clinical trial process and, with more money than ever in pharmaceuticals, pulling down that edifice is not for the faint-hearted. Though Colloca and Benedetti wrote that these revolutions in our understanding of placebo "will lead to fundamental insights into human biology," it is surely in this radical overhaul of medicine that the anomaly of placebo will create a Kuhnian paradigm shift.

Testing drugs has progressed enormously since Franklin's day. The modern apogee is the *randomized controlled trial* (*RCT*), where a large group of people is split into (usually two) groups on an entirely random basis. One group will receive the drug; the other will receive something that seems the same but is entirely inert: the placebo. The idea of randomization is to create as little natural difference between the groups as possible, thus maximizing the chances of seeing some effect the drug produces that the placebo doesn't. Systematic effects, such as gender, age, preexisting health issues, a natural swing into or out of good health, should be the same for both groups. Any major differences in outcome between the groups, then, should be due to the drug.

There are other factors at work, though, which is where blinding comes in. Obviously, none of the patients should know whether they are getting the drug under test or the placebo. This single blinding isn't enough; the people giving out the drugs might offer some nonverbal or subconscious clues to the patients. Hence the "double-blinding": the doctors and nurses involved also ought not to know which are the placebo pills.

Such a double-blinded RCT is considered the best way to tell whether a drug is effective or not, but there are still more refinements that can improve things. Adding a third "arm" to the study—a group that receives no treatment whatsoever—can help. Patients are most likely to seek a doctor's help when their symptoms are most acute; any follow-up is likely to encounter improvements in health. A group that has received no treatment will help weed out this "regression to the mean" effect. Similarly, there is the problem of "natural history": the normal variation in symptoms. A headache comes and goes, for example; if a patient takes a placebo just before a spontaneous swing toward less pain happens, the reporting could end up skewed. Observing a no-treatment control group should enable this effect to be taken into account.

Nevertheless, there are subtle effects that no amount of care seems to nullify. Just telling patients they *might* get a placebo alters the outcome. Telling them the likely potency of the drug will also skew things. A patient's own assessment of whether he is in the placebo or the active arm of the trial affects his response; two trials—one in Parkinson's patients, one in acupuncture—have been reported where the "perceived assignment" had more effect on the patients than the treatment on offer.

Because of all these factors (and there are others), the National Institutes of Health is sponsoring many different research groups to find a new way to test the efficacy of drugs. One group, led by researchers from Harvard Medical School, are attempting a new style of trial using "wait lists" to give them a control group that receives no treatment. Another way forward is through hidden treatments: covert versus overt treatment. The level of placebo response—and thus the effectiveness of the drug—can be determined by the difference in outcome between the group that knew they were getting the drug and the group that didn't know they were getting it.

So far, these trials have provided rather striking outcomes. An openly administered dose of the painkiller Metamizol, for instance, relieved postoperative pain much better than a hidden dose; all of the open-administration group's relief was from placebo. When researchers injected a different set of patients with a hidden dose of the painkiller buprenorphine, this did have a pain-reducing effect—though not as much, or as fast, as giving it through an overt injection. Though buprenorphine works, it works better when used in conjunction with the placebo effect. This kind of trial, which allows physicians to see the total effect of drug plus placebo, can help them give a reduced dose of potentially toxic or addictive substances.

Skeptics might argue that pharmaceutical companies will fight anything that casts their products in a dubious light—especially if it results in people using lower doses across the board—but the truth is that, for many drug companies, reliable information on the placebo effect can't come soon enough. To pass muster, a drug must outperform placebo. But a 2001 study of antidepressant drug trials showed that while drug efficacy is rising, placebo rates are rising faster. It's almost ironic; the factors behind this are many and varied, but a significant contributor is our society's knowledge of—and belief in—the power of medicines. The pharmaceutical industry's palpable success means that unless something radical happens, it could soon be, like the Red Queen, running to stand still.

The other big opportunity for a paradigm shift is in the clinical scenario: should we ignore Hróbjartsson and Gøtzsche and encourage doctors to keep lying to us about their treatments?

Health-care providers may not like the idea that the bold future of medicine lies in more exploitation of the healing power of the imagination, but if doctors are serious about preserving your health—maybe even saving your life—they might have to swallow this bitter pill. Not because placebo is a magic bullet, but for precisely the opposite reason. For all the marvels of the placebo effect, perhaps the most important thing is to recognize its limits. The placebo effect will not cure cancer. It does not slow the onset of Alzheimer's or Parkinson's. It does not make a malfunctioning kidney function again. It does not protect against malaria. Patients are already flocking to "complementary" therapists who unwittingly embrace placebos. The same patients are probably unaware that their family doctor could quite intentionally—some would say "cynically"—embrace these same treatments too, where appropriate.

It might be a disaster if they don't. The danger comes when the complement part of complementary disappears, and patients visit practitioners offering only "alternative" treatments. If the patient's condition is simply not placebo responsive—even if many of the symptoms are—that could be life-threatening. Get the placebo out in the open, find a way to make it an acknowledged tool in the doctor's armory, and we could save lives by keeping patients within the fold of efficacious, rational medicine. Just as long as we admit that, for the moment at least, it's not quite as rational as we'd like.

And that brings us to our last subject. It is, to many minds, not qualified to stand alongside these others. However, we have just raised questions about the placebo effect and the clinical trial, and these both have a bearing on the claims made for science's least favorite anomaly: homeopathy.

HOMEOPATHY

It's patently absurd,
so why won't it go away?

An insightful mind once remarked that historians labor under a delusion: they think they are describing the past, when in fact they are explaining the present. It must be doubly true of the historians of science. Time and again, going through these anomalies, we have had to dig into history in order to understand what is happening in contemporary science, and where its future might lie. With our final anomaly, it turns out that the insight is particularly powerful.

Homeopathy, invented in the late 1700s, is now more popular than ever. According to the World Health Organization, it now forms an integral part of the national health-care systems of a huge swath of countries including Germany, the United Kingdom, India, Pakistan, Sri Lanka, and Mexico. The United Kingdom's homeopathic hospitals, part of the country's national health service, treat around fifty thousand patients a year. Forty percent of French physicians use homeopathy, as do 40 percent of Dutch, 37 percent of British, and 20 percent of German physicians. In 1999 a survey revealed that 6 million Americans had used homeopathic treatments in the previous twelve months. The big question is, why? An assessment of homeopathy using the criteria of known scientific phenomena says it simply *cannot* work; no wonder Sir John Forbes, the physician to Queen Victoria's household, called it "an outrage to human reason."

Although there are several different approaches, homeopathy generally involves first finding a cure by the *principle of similars*, which says that the remedy should be of a substance known to create the very symptoms the patient is already suffering. Then that remedy is diluted in water or alcohol to the point where the solution handed to the patient contains no molecules of the original remedy. Nonetheless, it has been "potentized" by repeated shaking or banging with each dilution—a process known as *succussion*. In fact, homeopaths say, this ultradilute solution is more potent in curing ailments than the original undiluted substance.

It sounds like a ridiculous idea and, to most scientific minds, it is. The statistics of dilution make it plain why. A typical homeopathic dilution is done in ratios of one part of the substance to ninety-nine parts alcohol or water (depending on whether the substance is soluble in water). This process is repeated—a dilution of one part of the original solution to ninety-nine parts water or alcohol—again and again. It's quite normal to do this thirty times—this is called a *30C dilution*. That means, if you started by dissolving a tiny amount of your remedy in around fifteen drops of water, you would end up with the original substance diluted in a volume of water fifty times bigger than the Earth. The big scientific problem with this is that when the homeopathic pharmacist sells you a few milliliters of this remedy, the math of chemistry tells you there is virtually no chance that it contains a single molecule of the original substance.

If you know the weight of a sample of some chemical—let's say carbon—the basics of high school chemistry tell you how many atoms you have in your sample. A gram of carbon, for instance, contains 5×10^{22} atoms. That sounds like a lot—and it is: it's 5 followed by twenty-two zeroes. In a 30C homeopathic dilution, however, there's not a lot left; if you take fifteen drops of liquid, you'll have no more than one ten-millionth of an atom. And since you can't split the carbon atom up (at least, not this way), it's safe to say you've got no carbon in there. In standard practice, medicinal effects come through interaction with the body's biochemistry, which means you need molecules of the remedy to be present in the body. With homeopathy, there's nothing. By any laws known to science, the remedy cannot interact with the biochemistry of your body in any meaningful way.

Samuel Hahnemann, the founding father of homeopathy, knew this, though; it's not about chemistry, he said, but about the "energy" of the remedy being passed into the water. Since this "energy" is not known to science, the obvious conclusion is that if a homeopathic remedy has an effect, it can be no better than placebo.

The first scientific counter to this point of view came from the laboratory of French immunologist Jacques Benveniste. In 1988 Benveniste convinced the journal *Nature* to publish the details of an experiment that showed water was permanently altered by molecules that had once been dissolved in it. The publication was on the condition that a rerun of the experiments be carried out in independent laboratories. That was done, in Marseille, Milan, Toronto, and Tel Aviv. After publication (with disclaimers), *Nature* requested that the experiments be done again, this time in the presence (and under the intense scrutiny) of three independent witnesses. *Nature*'s then-editor John Maddox, the magician and professional skeptic James Randi, and Walter Stewart, a chemist and an expert on scientific fraud, spent a week in Benveniste's Paris lab. The full tale is an extraordinary one; the short version is simply that the visitors discovered how Benveniste had been duped by his assistant, who was cherry-picking data to support her belief in homeopathic medicine.

Nature published a critique of the original paper. Benveniste fought back, citing a McCarthy-like witch hunt, but his goose was cooked. The following year, his employer, the French National Institute of Health, criticized him for credulousness, cavalier reporting of his results, and abuse of his scientific authority. Two years after the *Nature* fiasco began, Benveniste was sacked.

That, essentially, was that—until Madeleine Ennis got involved. Ennis, a professor of immunology at Queens University, Belfast, says she was a hard-nosed skeptic of homeopathy and the Benveniste work. When she expressed this in the face of a published homeopathic trial, a manufacturer of homeopathic remedies asked her to join a team that would make another attempt to replicate that result. She agreed, expecting to add to the evidence against homeopathy. After the end of the trial, she declared herself "incredibly surprised" by the result. Quoted in the *Guardian*, she says, "Despite my reservations against the science of homeopathy, the results compel me to suspend my disbelief and to start searching for a rational explanation for our findings."

The trial, which was essentially a replication of Benveniste's experiment, took place in four different laboratories in Italy, Belgium, France, and Holland. Ennis's skepticism wasn't the only safeguard. The homeopathic solutions (and the controls) were prepared by three independent laboratories that made no other contribution to the trial. Inside those solutions were— or rather, had been—molecules of histamine.

Anyone who suffers from hayfever knows the power of histamine: it's an immune system response that produces hives, pain, itching, swelling, constriction of breathing, runny nose, and streaming eyes. All that, from some tiny molecules that form a small part of your bloodstream. Every drop of blood contains somewhere in the vicinity of 15,000 white blood cells; around 150 of those cells are known as *basophils*, and inside these basophils, contained in tiny granules, is the histamine.

Histamine has a strong effect on its basophil containers. After they release the histamine, its presence in their environment stops them from releasing any more. This effect was central to Ennis's experiment.

The labs that prepared the ultradilute histamine solutions sent test tubes of water and test tubes of dilute histamine to the labs carrying out the experiment. The histamine dilution was at the kind of level homeopaths routinely use, where there would have been no molecules of the substance in the vials. There was no way to tell which was water and which was the homeopathic solution. In the experiment, the researchers stained basophil granules blue, then put these colored granules into the test tubes, along with a substance called *anti-immunoglobulin E* or *algE*. AlgE causes a *degranulation* reaction, in which the color disappears and the granules release histamine.

In water, this is exactly what happened. But when the researchers put the colored granules and the algE into the ultradilute histamine solution, the degranulation didn't happen. The "ghost" presence of histamine in the homeopathic solution was enough to stop the process in its tracks.

The results were statistically significant at three of the centers. The fourth center had a positive result: the histamine solution did suppress de-granulation more than the pure water, but it was not different enough to count.

Ennis was not satisfied by the results; there could have been bias in iden-tifying which basophils still had their blue color because the researchers did it by eye. So she demanded they make a different measurement, one that could be automated. That way, a believer among them would not be able to skew the results—even unconsciously. She had the basophils "tagged" with an antibody that would make them glow if their histamine secretion was be-ing suppressed. A light-sensitive probe then did the counting. The result was the same.

The record of the experiment, published in *Inflammation Research*, con-cluded that "histamine solutions, both at pharmacological concentrations and diluted out of existence, lead to statistically significant inhibition of ba-sophil activation by anti-immunoglobulin E."

Not that Ennis quite puts her own results beyond question. It was, she admits, a small study, and no one has yet replicated its findings. In one fa-mous attempt, a team of scientists failed to replicate Ennis's experiment for a BBC *Horizon* program. Ennis appeared on the show, but she later dis-tanced herself from the experiment, saying there were a series of flaws in the protocol. A study by Adrian Guggisberg and colleagues at the University of Bern also failed to find any effects from homeopathic histamine dilutions. The Swiss team's analysis of protocols and results, published in *Complemen-tary Therapies in Medicine* in 2005, found that small variations in the exper-imental setup could lead to significantly different outcomes; there were all kinds of things that could affect the experiment, such as the temperature at which the basophils were prepared, and how long in advance the homeo-pathic solutions were prepared.

Homeopaths will certainly cry "aha" at one of the Bern study's main ob-servations: the results "might depend on inter-individual differences of blood donors," according to the paper's conclusions. The idea that homeo-pathy works on a case-by-case basis, that a remedy will produce healing ef-fects in some people and not in others, has been the homeopath's primary excuse when confronted with negative results in clinical trials of homeo-pathic remedies. Almost every time a homeopathic medicine fails to register an effect, a representative of homeopathy will respond by saying homeo-pathic prescription is a complex process; symptoms have to be considered in the light of all other aspects of the personality and physiology, and the right remedy for an ailment will be dependent on a large number of factors. Ask a homeopath to prescribe for an ear infection, say, and she'll ask, Which ear? Since the body isn't symmetrical—the liver and the heart, for example, lie away from the center line and, unlike the kidneys, have no mirror organ—ailments affecting one side of the body will have a different nature from ail-ments affecting the other. Even if your two ears do look the same.

To a scientific mind, that just comes across as an untestable waffle. Which is why, in the end, almost every scientific mind says homeopathy can't work. Even when that scientific mind acknowledges evidence to the contrary does seem to exist.

In his book *Placebo*, Dylan Evans attributes any success of homeopathy to the placebo effect. However, he also admits that a 1997 meta-analysis published in the *Lancet* shows it is, on average, significantly more effective than a placebo. How does Evans square this circle? By saying that "it would be foolish indeed to cast aside the whole of physics, chemistry and biology—supported, as they are, by millions of experiments and observations—just because a single study yields a result that conflicts with their principles." The University of Maryland skeptic Robert L. Park uses the same argument. "If the infinite-dilution concept held up, it would force a re-examination of the very foundations of science," he says.

Is this true? If ultradilute solutions can have effects on biology, will this send science back to the drawing board? No. Science works; millions of ex-periments and observations can be explained using scientific principles. None of those results are changed if homeopathy turns out to be right. Why? Because none of those millions of experiments and observations has told us everything we would like to know about the microscopic properties of water.

WE know very little about liquids. Solids are easy; for decades it has been possible to probe the structure of solids using techniques such as X-ray dif-fraction. That is how Francis Crick, James Watson, and Rosalind Franklin worked out the structure of DNA; they bounced X-rays off the crystal and interpreted the resulting regular X-ray pattern to reveal its regular arrangement of atoms. They key word here, however, is *regular*. Liquids aren't regular, and we have no way of probing an irregular microscopic structure.

Chemists assume that in the absence of external influences, the structure is likely to be similar all through a liquid; the chemical bonds should surely arrange themselves so there's minimum stress in the setup. But what happens at fluctuating temperatures? Or if there are regions of the liquid under high pressure? Or in electromagnetic fields? Can water in a jug exist in fairly neat order in some regions and clumped messily in others? Does it interact with the molecules in the glass walls of the jug? We don't know.

One thing we do know is that water is a particularly strange liquid. A stone's throw from the brown ooze of the River Thames, across from the Houses of Parliament, is the office of a man who could be considered the world expert on water. Martin Chaplin, a professor at London's South Bank University, has dedicated his career to studying the wet stuff and its scientific properties. How many anomalies does it show? At least sixty-four, he says.

Most of that weirdness comes from the weak bonds that exist between water molecules. The oxygen atom in H_2O has a couple of electrons that are not engaged in bonding to the hydrogen atoms. Their negative charges are, however, attracted to the positive charge in the hydrogen atoms of other water molecules.

Though these bonds, known as hydrogen bonds, are weak—at room temperature they are constantly being broken and re-formed as the molecules slide around each other—they are responsible for many of water's strange properties. In fact, they are responsible for your existence; water's hydrogen bonds are what makes Earth habitable for humans. The hydrogen bonds make water the only liquid that expands on freezing, for example. That means ice doesn't sink to the bottom of an ocean; if water were like every other liquid in this regard, the planet's oceans would be frozen solid, with only the top layer melted by sunlight. Complex life would be untenable.

Water's properties also lie more directly behind the phenomenon we call *life*. When one of the *Nature* journals asked Chaplin to write a review of water's role in biology, he started it with a rather provocative statement. "It is surely time," he said, "for water to take up its rightful position as the most important and active of all biological molecules."

Chaplin is the campaign coordinator, the chief of staff, for the recognition of water's role in our world. His review article reads like a political address. Studying other, "glitzier" biomolecules might be fashionable, but water is the key to all of them, he says. When the proteins, the workhorses of your body, fold up to take on particular shapes and roles, water, thanks to the electrostatic attractions its hydrogen bonds provide, is a necessary part of the process. And then, when a protein has finished forming, water is the protein's lubricant, its hydrogen bonds allowing the protein to flex as it goes about its business. Water is as essential to a protein as the amino acids that make up the protein chain.

In DNA, water molecules form electrostatic links with the base pairs; the orientation of the water molecules varies with the bases and the sequence they are in. It is this pattern of water molecules, and its resulting electric field, that allows proteins (with their own water) to approach and bond with the correct base pairs—and to do it quickly and accurately. Thus water is essential to processing the information contained in the DNA; it is at the center of the phenomenon of life. "Liquid water is not a 'bit player' in the theater of life—it's the headline act," Chaplin says. "Water can function as individual isolated molecules, small clusters, much larger networks or as liquid phases that can have different 'personalities.'"

In 1998, for instance, Chaplin was working out how the attractions between the molecules would cause water molecules to form groups. His calculations showed that water could well exist in 280-molecule clusters that took the shape of a twenty-sided solid where every face is an equilateral triangle. We know this shape as an icosahedron. Buckminster Fuller took it as the basis for his geodesic designs, but we also see it in nature; many viruses adopt the shape because it is the most efficient way to pack their proteins.

Interestingly, the shape has an ancient connection with water. The Greek philosopher Plato identified five "perfect solids," which he associated with elements and aspects of the universe. The cube he called Earth; the tetrahedron, Fire; the octahedron, Air; the dodecahedron, the Cosmos. Water, to Plato, was the icosahedron. Which makes it all the more surprising that in

2001, three years after Chaplin first suggested water might take this form, a group of German researchers saw the shape in a tiny drop of water around a millionth of a millimeter across.

The icosahedron is just one of many ways in which water molecules can cluster; there are pentamers, octomers, decamers, ice-seven, and hexagonal ice . . . and that is only one aspect of the structure in water. In 2004 Tatsuhiko Kawamoto and his colleagues published a paper in the *Journal of Chemical Physics* showing that as you squeeze or cool a body of water, it becomes broken up into distinct beads, each of which has characteristics slightly different from surrounding beads. It's almost like a pebbled beach; from a distance the shore looks smooth and continuous, but when you jump off the promenade, you find yourself walking on stones of varied color, roughness, shape, hardness, and size. The origin of all these differences in water, Kawamoto found, was in the hydrogen bonds that weakly link water molecules to each other. Each of these bonds responds in a different way to pressure; just as pebbles get eroded at different rates and in different ways by the waves crashing on a shore, the hydrogen bonds in a body of water will respond individually. The result is a rich mess of water "aggregates."

Further evidence of the heterogeneous nature of water came in 2004, when scientists led by the Stanford physical chemist Anders Nilsson published a paper in *Science* showing that water could exist in chains and rings. Water is far more interesting than just a sea of identical molecules of H_2O. In fact, it seems naive, in the light of the evidence that research has thrown up, to think of water as composed of just plain water molecules.

NOT that any of this is a proof for homeopathy. Most scientists are reluctant to get involved with explaining homeopathy via the structure of water. The field has been tainted since Benveniste's announcement and subsequent fall from grace. One might say he is both the Pons and the Fleischmann of homeopathy, and nobody wants to share his fate. In fact, the parallel goes farther, because any ideas people have about how the complexities of water might explain the claims of homeopathy are as unsatisfying as the theories that attempt to explain cold fusion.

Nevertheless, there have been attempts to explain how homeopathy might work. Perhaps the best offering so far came in a paper published in *Materials Research Innovations* in 2005. At first glance, the four authors certainly make an impressive lineup: Rustum Roy, the founding director of the Pennsylvania State University's Materials Research Laboratory; M. Richard Hoover, an assistant professor also at PSU; William Tiller, a former department chair of materials at Stanford University; and Iris Bell, a professor of medicine, psychiatry, family and community medicine, and public health at the University of Arizona.

Most of the paper is a literature review. It points out that the structure of a material, not its composition, controls its properties. The distinction between the different forms of carbon—graphite is a soft lubricant while diamond a hard solid—makes this point rather conveniently. In water, many structures exist (they cite Martin Chaplin's observations that water has been seen to exist in clusters composed of anywhere from 2 to 280 molecules), suggesting the potential for many differing properties to emerge within one body of liquid. The authors point out that of all liquids and solids, water moves between its different structures with the most ease.

Perhaps the most compelling point in the paper, though, is the discussion of *epitaxy*. Epitaxy is a well-known phenomenon in which structural information is transferred from one material to another without the transfer of material or the involvement of chemical reactions. The way some wafers of silicon are grown in the semiconductor industry offers an example. Place a solid crystal—often a lump of gallium arsenide, but it could be glass or ceramic—in a solution of silicon dissolved in liquid gallium. By controlling the temperature conditions, you can make the silicon slowly come out of solution and deposit its atoms on top of the crystal. The way it grows—that is, where its atoms fall and how the lattice structure forms as each atom comes out of solution—is determined by the structure of the outer layers of the original substrate crystal. The spacing of the substrate's atoms and the orientation of its lattice structure will, effectively, dictate how the new silicon crystal forms. This process is known as *liquid phase epitaxy*, but deposition from vapor, or even a beam of vaporized material, is also widely used in semiconductor manufacturing. If you have a computer, a pacemaker, or a high-tech toaster, the chances are that at least one of its components was made using epitaxy.

Rustum Roy and his colleagues make the point that the original material of a homeopathic remedy placed in water might have a similar epitaxial effect on the water (or water plus ethanol) of a homeopathic dilution, altering its structure. That altered structure could then be passed on as the solution is further diluted—especially with succussion. The "imprinting" of structure, they suggest, might be made possible by the high pressures created in the succussion process. Since it is structure, not composition, that determines properties, the absence of molecules of the original remedy in the final solution is thus immaterial.

As far as it goes, the array of possible mechanisms for the "memory of water" is intriguing. It is unfortunate, then, that Roy and his coauthors didn't refrain from examining the effects of electromagnetic fields and human intention, which they refer to as "subtle energies," on water when they wrote their paper; it rather has the effect of breaking their spell.

The team of researchers involved in putting this paper together might have impressive academic backgrounds, but there are also reasons to take what they say with a little pinch of salt. With the exception of M. Richard Hoover, they have reasons, besides the open-mindedness of a scientific approach, to want homeopathy taken seriously.

Roy, for example, has a long list of emeritus professorships, and an even longer list of publications in respected journals. He received a research award from the emperor of Japan; he even had a mineral—Rustumite—named after him. On the down side, though, Roy associates professionally with Deepak Chopra, whose claims for the healing quantum properties of water are questionable, to say the least. Roy advocates using silver as an antibiotic, something that has repeatedly separated fools and their money—including those selling the silver, who have been fined by the FDA for promoting and profiting from a treatment that can result in actual bodily harm. He also thinks—and advocates in this paper—that the conscious will of a healer, such as a Chinese Qigong grand master, can change the structure of water. Tiller, for his part, has published claims that weak magnetic fields can alter biological materials and the pH of water, and that human intention can also change pH, affect electrical circuits, and alter the properties of space. Iris Bell is an enthusiastic advocate of holistic and alternative medical practice—a lesser problem, but still worth noting.

Despite such damning caveats, the paper does make some truly interesting and potentially important points, offering hints as to where further research might clarify our understanding of the mechanisms that could lie behind homeopathy. The question is, will anyone want to pursue them? Is homeopathy worth our attention?

Evidently, millions of people think so, judging by the uptake of homeopathy. There is also the fact that it is absorbing public money to consider. Some scientists, Richard Dawkins, for example, are vociferously up in arms about the idea that their taxes are funding this "quackery." Are they right to be outraged? Answering that depends on the answer to another question: does homeopathy work or not? If only the answer were as simple as the question.

ON August 27, 2005, the *Lancet* announced the "End of Homeopathy." Its editorial article said homeopathy could no longer make any claims of efficacy, and that doctors "need to be bold and honest with their patients about homeopathy's lack of benefits." The reason for the declaration was an article published in the same issue, a meta-analysis of homeopathy led by Aijing Shang of the University of Bern and published to great fanfare. It pronounced homeopathy "no better than placebo." Since we have already discovered that meta-analyses of placebo trials have declared the placebo effect quite possibly a myth, that perhaps doesn't seem so ground-shaking. But, for Shang and his team, the study provided a killer blow: homeopathy was dead.

Until about a week later, that is, when the letters started coming in.

Although the authors had claimed their analysis put the final nail in the coffin of homeopathy, some scientists—not just the friends of homeopathy, it has to be said—were appalled that the *Lancet* had published such a "flawed" study. Klaus Linde and Wayne Jonas had published a very similar analysis of the medical literature about homeopathy in 1997—also in the *Lancet*—and felt compelled to complain. "We agree that homeopathy is highly implausible and that the evidence from placebo-controlled trials is not robust," they said. "However, there are major problems with the way Shang and colleagues presented and discussed their results, as well as how the *Lancet* reviewed and interpreted this study."

For a start, they pointed out, Shang's group did not follow the accepted guidelines for reporting meta-analyses. They left out details of the trials they examined, and excluded details of the trials they had decided to leave out of the review. In a paper that came to such a strong and definite conclusion, such lack of detail was "unacceptable," Linde and Jonas said. By its own standards, declared in 1999, the *Lancet* should have refused to publish the study.

The other big problem was that Shang's study involved pooling data from trials measuring different effects: different kinds of homeopathic treatments for different kinds of ailments with different kinds of outcomes— pain reduction, resolution of an infection, reduction of inflammation, and so on. That is OK if homeopathy really is nothing more than a placebo, because all trials are, effectively, measuring one kind of response. Linde and Jonas's 1997 study had pooled data on this assumption. But since then several studies had found some effect above placebo in trials of homeopathy in specific cases. If those studies have any truth to their results, Shang's pooling of results invalidates the entire analysis; it skews the statistics, giving a significant risk of producing a false negative.

Lastly, once Shang and colleagues had whittled down the trials they decided were worthy of attention, this meta-analysis ended up studying only eight clinical trials of homeopathy. With such a small pool, the outcome "could easily be due to chance," Linde and Jonas said. And that means their suggestion that they had proved the clinical effects of homeopathy are placebo is "a significant overstatement."

In 1997 Linde and Jonas had concluded that the results of their own study made it impossible to claim that the effects of homeopathy are completely due to placebo. It was hardly a ringing endorsement, but neither was it a nail in homeopathy's coffin. In fact, they lamented in their letter to the *Lancet* that their own study had been "misused by homeopaths as evidence that their therapy is proven." The *Lancet*, they said, was misusing the new study in a similar way. "The *Lancet* should be embarrassed by the Editorial that accompanied the study," Linde and Jonas said. "A subversive philosophy serves neither science nor patients."

Strong words, especially since they came from nonbelievers. But then Jonas had been encountering his own frustrations with homeopathy. A couple of months later, in October 2005, he published a paper with Harald

Walach in the *Journal of Alternative and Complementary Medicine*. It is a balanced review, admitting that there are "some hints" that diluted and succussed homeopathic substances are biologically active—but that there is "no single clinical area where reported effects have been demonstrated unequivocally." Meta-analysis, they say, just doesn't help; whether homeopathy comes out as a placebo or not depends on the way a study is done. Overall, the problem with analyzing homeopathy's credentials "is not in finding a stunning initial result . . . the real problem is replicating the effect once it has been seen." In other words, they too had failed to prove homeopathy's inefficacy. Yet again.

This all seems implausible. Given more than two centuries, science has failed to show that homeopathy is bunkum. How is this possible? How can we put this issue to rest? The answer may lie in the pages of the *Homeopathic Repertory*, the sprawling catalog of symptoms, remedies, and appropriate dilutions consulted by homeopaths before offering a prescription. The clinical trials that have tested homeopathy's efficacy pick out some homeopathic remedies and use them to treat ailments such as the inflammation caused by rheumatoid arthritis. A six-month study carried out by the director of research at London's Royal Homeopathic Hospital gave a negative result for forty-two homeopathic remedies in this case. But what if some of the remedies being used are in fact effective? Could it be that concentrating on just a few of the myriad available remedies would cause the results of trials to come out as significantly better than placebo?

It would certainly make sense of the gulf between the generally unimpressive trial results and the anecdotal claims that sane individuals make of homeopathy's successes. Lionel Milgrom, a chemist at Imperial College London, for example, trained as a homeopath because he was so impressed at how quickly and conclusively a homeopathic remedy cured his partner's recurrent pneumonia. Another acquaintance, an author of science books and a professional science communicator, once told me he watched in amazement when *Apis mel*, a homeopathic remedy made from a honeybee drowned in alcohol, deflated his two-year-old daughter's swollen tongue after she had been stung by a bee.

It might be that all these miracle stories could be properly balanced by reports of incidents where homeopathy fails to have an effect. It's called publication bias in the pharmaceutical trade; people seldom bother to announce exclusively negative results. But here's the rub: Could it be that the haphazard nature of homeopathic diagnosis and prescription—and the flakey, unproven (and yet unquestioned) status of some of its remedies—is occluding a truly impressive phenomenon?

Bob Lawrence certainly thinks so, and he's on a mission to prove it. Lawrence is another convert; he had a fifteen-year skin condition cleared up by a homeopathic remedy. Antibiotics also cleared it up, he says, but the side effects were too horrible to live with. When a friend recommended a homeopathic treatment, he took it with a large dose of skepticism, but he hasn't looked back. He subsequently gave up a perfectly good job in engineering to train as a homeopath and is now a pharmacist at one of Britain's largest homeopathic dispensaries, the Helios Homeopathic Pharmacy in the English spa town of Tunbridge Wells. Take a tour of this place with Lawrence, and you'll encounter everything that's wrong—and right—with homeopathy in the twenty-first century.

I was expecting something a little more disconcerting, something a little more *Romeo and Juliet*. Something like an apothecary's lair. Instead there's a brightly lit shop, a service counter, and behind that, a load of very normal-looking people. They were bustling about in white coats, sliding boxes off shelves, then opening them up and pulling out tiny vials from which they dripped liquids into other tiny vials.

There were three disconcerting things about the scene, though. The first was the strange names typewritten onto the box labels—one of them said "Lava," for instance. Also troubling was the occasional violent banging noise as a pharmacist succussed a vial's contents. The third oddity was the surface on which Lawrence was doing his own succussion. It was a huge, black, leather-bound King James Bible.

Having done three raps on the Bible, his fist clenched around a vial containing a homeopathic remedy made from amethyst, Lawrence looked up. His face said, "I wish you hadn't seen that." You don't have to use a Bible, he assured me. What you need, apparently, is a flexible but hard surface. It was Hahnemann, the founder of homeopathy, who suggested a leather-bound Bible might be the most suitable tool.

Hahnemann is followed enthusiastically at Helios. The pharmacy is a kind of central clearinghouse for homeopathic remedies; the staff are expert at the kind of dilution, succussion, and redilution that lies at the heart of homeopathy, and they get sent substances to "potentize" in this way from all over the world. The pharmacy's Bible has obviously seen a lot of action; its covers are now held on with rubber bands.

Lawrence is no mystic, though; he is not a pathological believer who thinks the Bible will convey some special force to the remedy. This was confirmed when he took me downstairs to show off the machines he built to do the most laborious succussions and dilutions. Sometimes, for an ultrapotent remedy, you have to repeat the process thousands of times; Lawrence has used his engineering skills to automate the process. He wants to put homeopathy on a more scientific footing, he says. Sometimes people will send him a bit of bat to potentize. Or the wing of a cicada. He won't turn it into a remedy until he knows exactly what species it came from; he wants the Latin name. He would dearly love to come at the *Homeopathic Repertory*, the sprawling catalog of symptoms, remedies, and appropriate dilutions, with a scientific ax, thinning it down to include only what's provably effective.

As we struggled to talk above the repetitive knocking of the machines, I noticed some more boxes. I could feel Lawrence willing me not to see the names on the labels, but his will is evidently not as powerful as that of a Chinese Qigong grand master. The names that stood out were "F Sharp Minor," "G Major Chord," "Crop Circle," and "Flapjack." When I asked Lawrence about them—how you capture F Sharp Minor in a bottle, for instance—he raised his eyebrows and rolled his eyes. Looking around again, I spied a box marked "Gog and Magog, Oaks at Glastonbury." Another said "Frog Spawn." Down here, in the basement, we had bypassed *Romeo and Juliet* and gone straight to *Macbeth*.

It's easy, in Helios's basement, to see what's wrong with homeopathy. It has, largely, become the preserve of people who want to believe in the healing power of anything and everything "natural." The range of homeopathic remedies is so vast, so all-encompassing, as to make it virtually impossible to test homeopathy's claims.

A homeopathic remedy is supposed to go through a system of checks known as a *proving*. The original substance is given to a group of volunteers, who note down any strangeness, any symptoms of anything that they experience over the next few weeks. These symptoms are compiled and compared, and the ones that seem universal are then associated with the substance. If a patient in a homeopathic consultation reports anything like those symptoms, the principle of homeopathy—literally, "similar suffering"—means that a remedy made from the substance in question might make a useful treatment.

The trouble is, many of the medicines in the Helios pharmacy have not had anything like proper provings—and are obvious examples of quackery. There are remedies on the shelves at Helios that have been made from condoms, pieces of lava, the blood from an HIV-positive man, and even the whiff of antimatter.

What's right about homeopathy, what it has going for it, is that someone like Lawrence is genuinely frustrated by this situation. I could see the embarrassment in Lawrence's eyes when I mentioned the musical remedies, and I feel a genuine empathy for his predicament. He says he doesn't have anything to do with these kinds of remedies, but he can't stop others from "potentizing" them. He believes homeopathy works, but he knows he doesn't have a clue why, and the extraneous, strange stuff on his shelves isn't helping anyone to find out. He wants to keep an empirical, almost scientific approach to the claims for homeopathy, while all around him flakes are making that nigh on impossible. Lawrence is actively trying to stem the tide of ridiculous remedies, but there is only so much that one man can do. Lawrence is not alone, however. Forty miles north of the Helios pharmacy, at the Natural History Museum in London, Vilma Bharatan is on a similar quest.

AS well as holding down a job as one of the Natural History Museum's botanists, Vilma Bharatan is a practicing homeopath. But she is also a fierce critic of homeopathy. Its practitioners, she says, have been living off people's reverence for the discipline without applying any intellectual or scientific rigor. They have been lax in laying out their data. The plant names they use are a mess, for instance, making it almost impossible to properly investigate relationships between known plant characteristics and reports of homeopathic usefulness. It wasn't always like this, she points out; there was a time when homeopathy was allied with science.

The pages of Bharatan's PhD thesis make interesting reading to anyone who wants to understand the problem with homeopathy. First they tie together a group of homeopathic remedies—the flowering plants—with their proper biological names, the symptoms they are meant to treat, and how effective homeopaths have found them to be. Then they sort these remedies using a computer program that performs a *cladistics* analysis.

Biologists use cladistics to group plants or animals according to their physical characteristics or their genetic profiles. Bharatan's plan was to load the program with the homeopath's idea of the therapeutic effects of the plants and see if there was any correlation between the homeopathic groupings and any of the traditional biological groupings.

Her database is called a *matrix*; it's a web of plant names, plus the claims for the various therapeutic effects each plant shows. Not that Bharatan included every homeopathic claim in her matrix; she restricted the data set to those that had, at the very least, been "frequently confirmed" in provings and their success confirmed in normal clinical use. In the end, the matrix comprised more than a quarter of a billion therapeutic effects of plant remedies. When she ran the data through the computer program that analyzed and sorted it, the museum's server was creaking under the strain. This was the largest data set it had ever analyzed.

The output from a cladistics program is called a *cladogram*. It looks like a kind of family tree. The cladogram that shows how insects evolved into their various forms, for example, branches off first for beetles. The other branch then splits into one branch for ants, bees, and wasps and another that branches into two again: one branch is the butterflies and moths, the other branch is the flies. From this picture, we see how recently two species descended from a common ancestor.

Bharatan's cladogram showed very few "common ancestors" for the most part; in many cases, the program found no strong biological relationships between the various homeopathic plant-based remedies. But on occasion it found very strong relationships. One grouping, or *clade*, branching off and subdividing rather like the insect group on the tree of life, contained remedies whose curative properties were associated with the cardiovascular system. Another group was plants used in treatments of female reproductive disorders. Look at the raw data for a million years, and you would never see these groupings, Bharatan reckons. Because the plants are used in such a wide variety of treatments, there is no way you would normally think of grouping them according to systems of the human body. Nor do they belong to the same botanical family. Nonetheless, after running for thirty-two hours solid, the computer decided they belonged together. The reason, it seems, is chemical.

If you are unfortunate enough to suffer congestive heart failure or an arrhythmia, your doctor may well prescribe you drugs containing cardiac glycosides. These compounds, which affect the way sodium and potassium ions move around in heart tissue, are usually derived from plants. Four of those plants, including *Digitalis purpurea*, the most widely used of the cardiac glycosides, are conspicuous by their presence in Bharatan's cardiovascular clade. In fact, all thirteen plants in the clade contain chemicals that are used in Western medicine for the treatment of heart-related problems: angina, heart pain, and irregular heartbeats, for example. Some of the chemicals reduce blood cholesterol, some slow the heart's contractions—there are all kinds of effects.

There are plenty of implications here, Bharatan says. First, the fact that the cladistics program found a pattern associated with systems of the human body challenges the idea that homeopathy works through placebo. If it is just a placebo, it's not clear where the pattern would come from, she points out. Second is the fact that, in Bharatan's analysis, plenty of plants came up as "noise" in the data; they weren't associated with anything useful, despite being in the homeopathic repertory. The cardiovascular clade, for example, did not include twenty-seven plants that were present in the matrix and are commonly used to treat symptoms of the cardiovascular system. Some of them, such as the tobacco plant, have a major effect on the heart; somehow, though, the computer decided they were not part of this clade. It's only a preliminary result, but it's intriguing, and Bharatan thinks her analysis might provide a scientific means of thinning out the overblown repertory of substances used in homeopathic treatments.

Bharatan does not want to stop there, however. The third—and probably the most striking—inference of her work, she says, is the suggestion from the cladistics that these homeopathic substances might be exerting a chemical action. Which means that dilution and succussion—to most, the very essence of homeopathy—could be not just a waste of time but the root of homeopathy's problems. If its power lies in chemistry, there is no need to jump through the hoops of imprinting structure on liquids; Rustum Roy might be barking up the wrong tree.

The whole concept of dilution and succussion is certainly questionable, she says; no one knows where it came from. Originally, Hahnemann used undiluted doses of plant-based treatments but got unwanted side effects. That's when he started watering his remedies down and succussing them. "That's what we can't explain," Bharatan says. "How did he come to try that out?" In asking the question, Vilma Bharatan is echoing the past—and risking the rejection of her peers.

MORE than a century ago, thanks to his disdain for extreme dilution, his fearless winnowing of the homeopathic materia medica, and a strong desire to move homeopathy closer to allopathic medicine, Richard Hughes was dismissed by his colleagues as a "skunk."

The editor of the *Annals of the British Homeopathic Society*, Hughes was a hugely influential character who stirred up no end of controversy during his lifetime. He was the first to stand up to Hahnemann, questioning his methods and criticizing those who followed him without thinking. Hughes (and many other British homeopaths following his example) diluted their remedies far less. Hahnemann's rule that the thirtieth potency—diluted in the ratio 1:100 thirty times—should be used had fossilized homeopathy, Hughes said. Instead they used nothing more dilute than 6C—six of the 1:100 dilutions. That, it should be noted, still reduces the material substance of the remedy to only one part per trillion.

This move toward lower dilution was part of what motivated Hughes's seven-year undertaking to rewrite the homeopathic materia medica using only reliable evidence. Out went anything relying on reports from treatments with dilutions above 6C. Out went purely clinical reports; Hughes dismissed them almost as hearsay. Everything was to be based on provings or reports of poisonings. The result, the four volumes of the *Cyclopaedia of Drug Pathogenesy*, was Hughes's magnum opus, hailed on his death in 1902 as "a work without parallel" and one whose pages would be "even more frequently explored at the end of the twentieth century than at its beginning." It wasn't to be.

Hughes's work had threatened to blur the line between homeopathic and allopathic medicine. He had expressed a desire to establish an era where "the rivalry between 'homoeopathic' and 'allopathic' practitioners would no longer embitter doctors and perplex patients." It sounded ideal until, wide-eyed, he pointed out the consequences: homeopathy "would at once cease to exist as a separate body." This dangerous ideal, according to a 1985 article in the *British Homeopathic Journal*, is most likely what caused his "posthumous ostracism." Nobody likes the prospect of being absorbed into a bigger organism, and within a few years of Hughes's death, homeopathy had retreated from its connections with science and become a metaphysical, occasionally mystical discipline.

And yet the spirit of Richard Hughes lives on. His materia medica, with its reduced-dilution "material doses," is part of the input data in Vilma Bharatan's matrix, the same matrix that, in a cladistic analysis, suggests current homeopathic prescription needs a radical rewrite.

The history of homeopathy makes it clear that the present standoff between allopathic and homeopathic medicine is an artifact of the past, not an indication of a fundamental incompatibility. In all likelihood, the reason homeopathy won't go away is simple: there is something to its prescribing principle, the action of similars. If Hughes had had his way, all the surrounding mysticism and mumbo-jumbo, the enfeebling dilution, and the noise of succussion would have been stripped away over the last hundred years, and the essentials of that principle might have been incorporated into allopathic medicine. Drug companies happily use local traditional knowledge of the healing properties of plants to find starting points for the development of new medicines, and there is no reason to think they wouldn't take homeopathic remedies just as seriously—if they didn't come with what Hughes referred to as the "fancies and follies" that have attached themselves to the basic prescribing principles.

Vilma Bharatan should certainly keep a tight hold on her cladograms; they might one day be seen as the filter through which homeopathic medicine came in from the cold. The irony is that in the harsh light of scientific scrutiny, homeopathy's only chance for survival and dignity may lie in its willingness to die.

EPILOGUE

I am in Wiltshire, England, on a final journey. Tomorrow I will be meeting with Martin Fleischmann, one of the two chemists behind the 1989 cold fusion debacle. Tonight, though, I am lying on a hilltop and staring at the stars.

Immediately behind me is an Iron Age monument, the undulating peaks and troughs of an ancient fort. Its ditches and mounds were built seven hundred years before the birth of Christ. Just below me, invisible in the dark, is a relative newcomer to the landscape, a white horse that was carved into the chalk at the orders of Alfred the Great. No one is quite sure when that was done—probably a thousand years ago. My view upward affords me yet another historical sight: in between the creation of the fort and the creation of the white horse came the creation of the light streaming out from the belt of Orion. Though it is only now hitting my eye, the three stars that make up Orion's belt blasted out this light around fifteen hundred years ago. It has been travelling ever since. When Alfred ordered the horse carving—as a celebration of his victory over the Danes—that light was still 6,000 trillion miles away.

It's nice to be able to put a figure on it; it is a privilege to live in an era when we know how fast light travels. In fact, we are privileged just to know that it doesn't travel instantaneously across the universe. We take such knowledge for granted, but we shouldn't; it was hard won.

In 1676 an anomaly in the orbit of Io, Jupiter's innermost moon, led the astronomer Ole Roemer to make a very specific prediction. Io would appear from behind Jupiter at 5:37 p.m. on November 9, 1676, he said—and that would prove light travels with a finite speed. Roemer's mentor, Jean-Dominique Cassini, head of the Paris Observatory, rubbished the idea; light spread instantaneously, he said. His beliefs led him to a different prediction. According to Cassini, it would be 5:27 when Io appeared.

Io appeared at 5:37 and 49 seconds. On hearing of this, Cassini announced that the facts fit with the story he had presented. Although Cassini had made his (erroneous) prediction at a public gathering of scientists, not one of them demurred when he denied it; they all backed him up. Roemer had to wait fifty years to be vindicated; only after Cassini had died did scientists accept that the speed of light was finite.

In 1969 the astronomer J. Donald Fernie made a wry observation. He was writing about the decades it took for astronomers to spot an error that had been made early in the twentieth century. "The definitive study of the herd instincts of astronomers has yet to be written," Fernie said, "but there are times when we resemble nothing so much as a herd of antelope, heads down in tight formation, thundering with firm determination in a particular direction across the plain. At a given signal from the leader we whirl about, and, with equally firm determination, thunder off in a quite different direction, still in tight parallel formation."

The words came three centuries too late to be of comfort to Ole Roemer, but we should take note; this is how science works. Just as light travels with a finite speed as it moves across the cosmos, science progresses with more impediment than you might ever have thought. However, there is no fundamental law that imposes a speed limit on science, to be sure. It is simply the fact that human beings are involved.

There are several factors in play. Sometimes, for example, people just don't notice things. When Wilhelm Roentgen discovered X-rays, at least one other researcher had already seen them but not remarked upon the strange nature of his observation. Sometimes, on the other hand, the human mind knee-jerks against a radical new idea. After Roentgen made his announcement, Lord Kelvin pronounced X-rays to be an elaborate hoax. Only later, after he had seen the experimental evidence, did Kelvin back down.

If other people don't get in the way, circumstances will. In 1905 scientists weren't really worrying too much about how the universe worked. At the beginning of the twentieth century, the Western world was dominated by heavy industry and agriculture, and that was where researchers directed their efforts. So when a Swiss patent examiner came up with a startling theory about the nature of space and time, no one took any notice. In fact, the theory of relativity didn't even help Albert Einstein get a job. When he applied for a teaching position, he enclosed his published paper and still failed to get an interview. It is almost ironic: the publication that used the finite speed of light to revolutionize our view of the cosmos couldn't do anything to speed Einstein's passage out of the patent office in Bern.

Sometimes the obstacle is a scientist's own fear of the unknown. Henri Poincaré was closing in on the theory of relativity well before Einstein. All the evidence was in place because special relativity is the perfect explanation for the results of an experiment carried out in 1887 by Albert Michelson and Edward Morley. Unfortunately for Poincaré, he abandoned the research when he saw its implications for space and time: that time slows down and speeds up depending on the way something is moving through the universe. It was more than he could face.

Then, when all else fails to block progress, there is always the assumption that there is nothing new to discover. Albert Michelson provided the classic example a whole decade before Einstein made his breakthrough. "The more important fundamental laws and facts of physical science have all been discovered," Michelson wrote in 1894, "and these are now so firmly established that the possibility of their ever being supplanted in consequence of new discoveries is exceedingly remote." Six years earlier, the astronomer Simon Newcomb had said we are "probably nearing the limit of all we can know about astronomy."

This self-assured triumphalism is not just an ancient phenomenon. In 1996 the science writer John Horgan published a book called *The End of Science*. Within its pages, Horgan argued that science is, essentially, finished. We are near a final theory of physics, he said, and there is little left of interest to discover in biology. All that is left is a bit of *i*-dotting and *t*-crossing. From here on in, science is boring; it is about filling in the details.

When Horgan's book came out, it provoked great anger among scientists. Stephen Hawking called it "garbage." Stephen Jay Gould called it "idiotic." It was even alluded to during a Nobel Prize acceptance speech that year; holding his Nobel for Physics, David Lee announced that rumors of the death of science were "greatly exaggerated." And yet the book had a significant and lasting impact. Three years later, the Nobel laureate Phil Anderson coined the term *Horganism* to denote a corrosive pessimism about science's future.

I have gotten to know John Horgan a little bit over the last couple of years, since we met at Cambridge University in the summer of 2005. I hold enormous respect for him. But I too think he is wrong. Yes, we now know the speed of light thanks to Ole Roemer, and we know myriad other facts about the universe and how it works thanks to the incessant progress of science. But there is also plenty left to do—and I am not talking about the boring stuff.

Since leaving the Hotel Metropole in Brussels, I have investigated just thirteen of today's scientific anomalies. Some are more anomalous than others, but all cry out for explanations and further study. Some have yet to be taken seriously; others are perhaps taken too seriously. The astronomer Simon White has, for example, suggested that the astronomical efforts directed at solving the dark energy riddle are probably too large compared with the benefit they will most likely give. Occasionally, the anomalies point us toward acutely uncomfortable facts that no one wants to face—such as our delusion of free will. But, for all their diversity, their thrilling or disturbing natures, each and every case presents a wonderful opportunity for exploration and discovery. They will also, as did radioactivity and quantum theory, lead us to uncover anomalies as yet unseen; as George Bernard Shaw once pointed out, science never solves a problem without creating ten more.

The ancient light painting the dark canvas above me is testimony to the truth of Shaw's statement. Roemer solved the problem of Io's orbit by postulating a finite speed of light. And a finite speed of light opened up another cosmic problem—one whose solution seems to be opening up a thousand more.

The stars are huge thermonuclear explosions that send out light and heat in the form of packets of energy. Our Sun is a smaller, closer version that gives us a more direct experience of light and heat; unlike Orion, it is close enough to bring us some way toward its own temperature. Roughly nine minutes ago as I lie here, the Sun belched out a photon that is now warming someone in Australia. I click my fingers now, and another photon rockets off from the Sun toward some early morning walker on Bondi Beach. In nine minutes it will be there.

Here, in the finite speed of light, is an anomaly. Though there is a marked difference between the temperature on Bondi Beach and the chill here on an ancient English hillside, the universe as a whole is remarkably uniform. Wherever you go, it is all roughly the same temperature: about three degrees above absolute zero, the coldest temperature possible. Which, given a finite speed of light, doesn't make much sense.

Perhaps that doesn't seem too strange at first glance. After all, we're quite used to things being at the same temperature. I'm lying here on the grass, and my feet are the same temperature as my head. My back is slightly cold because the ground is leeching some heat from me, but essentially I'm the same temperature all over.

That's only true for the same reason as the stars shine, however: hot things emit radiation. The radiation carries energy in the form of photons that collide with other things—generally less hot things. Collisions transfer energy from the hot thing to the cold thing until they are both at the same temperature. Given enough time, things reach equilibrium.

The problem is, the universe hasn't had enough time to reach its equilibrium. There must have been all kinds of chaos just after the big bang; the universe was definitely not uniform at creation. And we know from various measurements of the stars that the universe is expanding, which means that in the 13.7 billion years since the big bang, the hurtling expansion of space has left some parts of the universe beyond the reach of others; the finite speed of light means the photons from the hot parts have not had time to reach enough of the cold parts to bring the universe to equilibrium. Yet everywhere we look, from horizon to horizon, the universe is pretty much exactly the same temperature.

Astronomers call it the *horizon problem*. Or rather they did until Alan

Guth solved it. Put simply, here's Guth's answer: just after the big bang, the universe blew up, very fast. Just like that. And it then stopped blowing up so fast and settled down to some respectable kind of expansion. For no reason that we yet understand.

It solves the horizon problem because before this period of ultrafast "inflation," the universe was small enough for photons to travel all the way across it, getting everything to the same temperature. Only after that had happened did the universe blow up.

No one knows how or why the universe might have started blowing up as Guth suggested. Or why the inflation suddenly stopped. It's hardly an explanation, really, but it is the best explanation we've got. Indeed, it is now so mainstream in cosmology, so unchallenged as a hypothesis, that you'd be forgiven for thinking that inflation was part of the well-documented history of the universe, somewhere just above the Battle of Waterloo on the scale of historically credible events. We may not know every detail of inflation, just as we don't know exactly how and when each of Wellington and Napoleon's soldiers died on that muddy Belgian field, but we now have good evidence that, just after the big bang, the universe did go through a phase of ultra-rapid expansion. It is a very neat solution to a very big problem.

Not everyone is convinced. Princeton's Paul Steinhardt doesn't think inflation happened, and the Nobel laureate Robert Laughlin, one of those pointing out the limits of reductionism, goes farther. The widespread acceptance of the standard ideas in cosmology—the big bang plus inflation—is unwarranted, he says, because scientists have adopted the *cosmic microwave background* radiation that fills all of space as the main supportive evidence. This radiation, sometimes known as the echo of the big bang, was generated three hundred thousand years after the beginning of the universe; the idea that it can tell us anything about the first few moments of creation "is like trying to infer the properties of atoms from the storm damage of a hurricane," Laughlin says.

Alan Guth solved a problem to most physicists' satisfaction. But Guth's triumph is really a doorway opened, and a new series of questions await us behind that doorway. They are not even difficult questions to generate, for the most part. Twenty-five years on, for instance, we are still stuck for the simple why and how of inflation. If the horizon problem was an anomaly, inflation is only a partial solution; really we have done little more than paper over our ignorance with an enigma.

The horizon problem is not, though, an anomaly I have explored in this book, partly because its explanation may well come from anomalies we have visited here. Investigations into dark energy or cold fusion or varying constants might bring us some deeper theory than quantum electrodynamics, for instance, and that new theory might play a role in explaining what could have caused the universe to inflate.

The solutions to the other anomalies might have similarly wide-ranging implications: investigating the origin of death and the story of the giant viruses might lead to radical revisions in evolution; understanding the placebo effect could—and probably should—change the face of medicine; coming to grips with the delusion of free will could alter the way we look at human beings and their responsibilities. It is safe to say, I think, that there is more than enough work ahead for the next generation of radical-thinking scientists—and the generation after that.

I chose to dedicate this book to the man who taught me physics when I was fifteen because the journey of discovery detailed within these pages ignited in me the same fascinations, the same passions that he ignited in me back then. Under his instruction, science became a thing of wonder, something to argue about, to explore, to provoke the mind. He taught me for little more than a couple of years, but he unearthed something in me that has lasted through more than two decades. And I might as easily have honored him by dedicating the book to his current students, to the next generation, the one that may solve these anomalies, creating many more in turn.

Kuhn observed that his paradigm shift model means that major discoveries are only made by people who are either very young or very new to that particular scientific discipline. Charles Darwin knew it too. In *On the Origin of Species*, he makes a telling statement. "I by no means expect to convince experienced naturalists whose minds are stocked with a multitude of facts all viewed, during a long course of years, from a point of view directly opposite to mine," he says. Instead, he adds, he is looking with confidence to the future, to "young and rising naturalists, who will be able to view both sides of the question with impartiality."

It will be the people who are now young and rising who will find life on the planets and moons of our solar system, maybe even answering a call from beyond those boundaries. It is they who will perhaps create life or rewrite Einstein's relativity to take account of dark matter and put the Pioneer probes to rest. Perhaps some genius still currently in preschool will use her mathematical skills to solve the riddle of dark energy.

Whatever the revolutions to come, one thing is sure. Every advance will most likely tell us as much about ourselves as it will about the universe we inhabit. We are collections of chemicals made in the cataclysmic explosions of stars; we are stardust, or nuclear waste, depending on your perspective. But, audaciously, we consider ourselves so much more than the sum of those parts; we declare ourselves to be alive, even though we don't know what that means. We want to, we expect to, discover other living things in this vast universe, while we also struggle to make sense of the chemistry of a few palladium atoms held in a small tank of water. We can think ourselves out of pain and yet can also prove we do not control even our own muscles. We launch probes into space, but we are unable to explain our most primitive urges and desires. We consider ourselves the pinnacle of evolution while aware we know very little of its true story. All this surely speaks to our desire to frame ourselves, to understand what it means to be a human being in this universe. And this is exactly what science—and the anomalies that drive it forward—can help us understand. "Who are we?" asked Erwin Schrödinger in 1951. "The answer to this question is not only one of the tasks but *the* task of science."

ACKNOWLEDGMENTS

It has been a privilege to write this book—I have never enjoyed anything more. In time-honored tradition, I must now thank all the people who allowed me to use their time, their labs, their colleagues, and their patience; the book couldn't have been written without them.

I would like to thank Fabrizio Benedetti, Luana Colloca, and Antonella Pollo for an extraordinary day in Turin; Patrick Haggard for a disturbing couple of hours in London; and the United States Navy's cold fusion researchers Pam Boss and Frank Gordon for their good humor when faced with difficult questions. I am grateful to Michael Melich and Martin Fleischmann for their insights over an entertaining (and delicious) lunch.

The list goes on: Gilbert Levin, a man of unusual dignity. Steen Rasmussen, a towering figure—physically and intellectually. Vera Rubin, an amazing scientist. The Pioneer researchers Michael Martin Nieto, Slava Turyshev, and John Anderson are also scientists of the highest caliber. John Webb and Michael Murphy are not just impressive and level-headed thinkers; they have always been great company too.

Jerry Ehman and Seth Shostak are due thanks for their candor about the hunt for intelligent aliens; Bernard La Scola, for giving me an excuse for a day trip to the sunny south of France; Joan Roughgarden, for helpful suggestions about sex; and homeopaths Melanie Oxley, Lionel Milgrom, Peter Fisher, and Vilma Bharatan, for their help with, and enthusiasm for, this whole project. I particularly enjoyed the company of Bob Lawrence, whose honest, down-to-earth approach to things that don't make sense gave me hope that solutions to the enigma of homeopathy might be possible. I also must thank Nancy Maret for her hospitality while I was in New Mexico.

I am grateful to Kris Puopolo of Doubleday and Andrew Franklin of Profile Books, both of whom gave enthusiastic support, excellent advice, and made extremely wise suggestions during the preparation of this book. My thanks also go to my agent, Peter Tallack of The Science Factory, who helped in myriad ways to get this book out of my head and onto the shelves. It wouldn't be right to leave my family out of the thank-list: my wife Phillippa and my children Millie and Zachary have put up with a distracted husband and father for long periods over the last couple of years.

Finally, during (and for years before) the writing of this book, I have gained enormous insight and clarity from discussions with my *New Scientist* colleagues: the collective brain of that magazine is an awesome organism. Jeremy Webb, Valerie Jamieson, Graham Lawton, Kate Douglas, and Clare Wilson were particularly helpful. Any mistakes in the text are their fault.

NOTES AND SOURCES

PROLOGUE

pp. 3–4   *he wanted to examine the nature of discovery*: T. Kuhn, *The Structure of Scientific Revolutions* (Chicago: University of Chicago Press, 1962), p. 10.

p. 5   *the U.S. Department of Energy recently declared*: Available at http://www .science.doe.gov/Sub/Newsroom/News_Releases/DOE-SC/2004/low_energy/ index.htm.

p. 5   *The philosopher Karl Popper once said*: K. Popper, *The Open Universe: An Argument for Indeterminism* (London: Hutchinson, 1992), p. 44.

1. THE MISSING UNIVERSE

p. 11   *Slipher is one of the unsung heroes of astronomy*: At the 207th Meeting of the American Astronomical Society, January 8–12, 2006, the Sonoma professor Joseph Tenn gave a talk titled "Why Does V. M. Slipher Get So Little Respect?" See also the Royal Observatory of Edinburgh cosmology professor John Peacock's Web site at http://www.roe.ac.uk/~jap/slipher/.

p. 11   *"probably made more fundamental discoveries"*: W. Hoyt, *Biographical Memoirs of the National Academy of Science* 52 (1980): 410.

p. 12   *Hawking makes a pointed reference*: S. Hawking, *The Universe in a Nutshell* (New York: Bantam, 2001), p. 76.

p. 12   *When these velocity measurements were published*: V. M. Slipher, *Proceedings of the American Philosophical Society* 56 (1917): 403.

p. 14   *the only explanation*: *Helvetica Physica Acta* 6 (1933): 110.

p. 14   *Dutch astronomer Jan Oort added to the evidence*: *Journal of the Royal Astronomical Society of Canada* 33 (1939): 201.

p. 16   *Cambridge professor Malcolm Longair . . . might turn out to be*: M. Longair, *Our Evolving Universe* (Cambridge: Cambridge University Press, 1996), p. 118.

p. 16   *Rubin published her results*: *Astrophysical Journal* 159 (1970): 379.

p. 17   *in 1999 . . . Rees gave an extension*: M. Rees, *Just Six Numbers* (London: Phoenix, 2000), p. 92.

p. 23   *The Harvard astronomer was worried*: R. Kirshner, *The Extravagant Universe* (Princeton: Princeton University Press, 2002), p. 192.

p. 25   *"This is nutty-sounding"*: K. Sawyer, "Cosmic Force May Be Acting Against Gravity," *Washington Post*, February 27, 1998.

p. 25   *"somewhere between amazement and horror"*: *Science* 279 (1998): 1298.

p. 25   *many of our finest minds seem to have given up*: *Nature* 448 (2007): 245.

p. 26   *Weinberg suggested . . . explain its value*: S. Weinberg, *Dreams of a Final Theory* (London: Hutchinson 1993), p. 177.

p. 27   *"unthinkable"*: L. Susskind, "A Universe Like No Other," *New Scientist*, November 1, 2003, p. 34.

p. 27   *Susskind calls them the Popperazzi*: A. Gefter, "Is String Theory in Trouble?" *New Scientist*, December 17, 2005, p. 48.

p. 28   *the physicists were similarly puzzled*: "Nobel Laureate Admits String Theory Is in Trouble," *New Scientist*, December 10, 2005, p. 6.

p. 29   *a characteristic feature*: *Astrophysical Journal* 523 (1999): L99.

p. 30   *As soon as Bekenstein developed*: *Physical Review D* 70 (2004): 083509.

p. 31   *"NASA finds direct proof of dark matter"*: See http://www.nasa.gov/home/hqnews/2006/aug/HQ_06297_CHANDRA_Dark_Matter.html.

p. 32   *There was nothing in the Chandra observations*: *Monthly Notices of the Royal Astronomical Society* 371 (2006): 138.

p. 32   *His modified gravity theory . . . any dark matter*: *Monthly Notices of the Royal Astronomical Society* 382 (2007): 29.

p. 33   *the Dark Energy Task Force issued their report*: See http://www-astro-theory.fnal.gov/events/detf.pdf.

p. 34   *hints that the universe is not isotropic*: *Physical Review D* 72 (2005): 101302(R).

2. THE PIONEER ANOMALY

p. 40   *In 2002 they published*: Physical Review D 65 (2002): 082004.

pp. 42   *Or maybe the signal photons . . . expansion of the universe?*: can be accessed on-
–43   line at www.arxiv.org/abs/gr=qc/0610034.

p. 43   *accelerated according to the laws of* nonlinear electrodynamics: *Europhysics
Letters* 77 (2007): 19001.

3. VARYING CONSTANTS

p. 48   *John Webb had what looked like an answer*: Physical Review Letters 82
(1999): 884.

p. 51   *His research team have dissected every result*: See, for example, *Physical Review
Letters* 95 (2005): 041301.

p. 53   *Their conclusion was probably disappointing to Dyson*: Nuclear Physics B 480
(1996): 37.

p. 53   *Steve Lamoreaux and Justin Torgerson . . . the energies involved*: Physical Review D 69 (2004): 121701(R).

p. 54   *a team of physicists published a paper*: Physical Review Letters 96 (2006):
151101.

p. 55   *Webb put the case for coolness like this*: J. Webb, "Are the Laws of Nature
Changing with Time?" *Physics World,* January 2001, p. 39.

p. 55   *Nobel Prize–winning physicist John Wheeler asked*: J. A. Wheeler, *Frontiers of
Time* (Amsterdam: North-Holland, 1979).

p. 56   *Feynman published a slim book on the theory*: R. Feynman, *QED: The Strange
Theory of Light and Matter* (Princeton University Press, 1988), p. 395.

4. COLD FUSION

p. 57   *a press release, issued on March 23, 1989*: Reprinted in J. K. Footlick, *Truth and
Consequences* (Phoenix: Oryx Press, 1997), p. 30.

p. 60   *The U.S. Department of Energy convened*: See http://www.ncas.org/erab/.

p. 60   *"as respectable in science as pornography in church"*: B. Daviss, "Reasonable
Doubt," *New Scientist,* March 29, 2003, p. 36.

p. 62   *In eight experiments*: Journal of Electro-Analytical Chemistry 296 (1990): 241.

p. 63   *Noting that Schwinger refused to follow*: Nature 370 (1994): 600.

p. 63   *"The pressure for conformity is enormous"*: From "Cold Fusion—Does It Have
a Future?" a talk given in Japan on December 7, 1991, at a celebration of

Shin'ichiro Tomonoga's centennial birthday. Available at: http://www/lenr-canr.org/acrobat/SchwingerJcoldfusiona.pdf.

p. 63   *Schwinger's attitude toward cold fusion*: "A Brief History of Mine," presented at the Fourth International Conference on Cold Fusion, Lahaina, Maui, December 6–9, 1993. Available at: http://www.infinite-energy.com/iemagazine/issue1/colfusthe.html.

p. 63   *"one of 20th-century physics' few unqualified triumphs"*: G. Johnson, "Two Sides to Every Science Story," *New York Times*, April, 9, 1989.

p. 64   *"Schwinger invited me to lunch"*: N. Ramsey, "Which Came First, Theory or Experiment?" *Physics Today*, January 2001, p. 13.

p. 64   *The journal duly published Schwinger's paper*: *Physical Review* 73 (1947): 416.

p. 64   *a Department of Energy study admitted*: See http://www.science.doe.gov/Sub/Newsroom/News_Releases/DOE-SC/2004/low_energy/CF_Final_120104.pdf.

p. 65   *an appendix added after publication*: S. Luckhardt, "Technical Appendix to D. Albagli *et al. Journal of Fusion Energy* article," *MIT PFC Technical Report (PFC/RR-92-7)*, discussed in E. Mallove, "MIT and Cold Fusion: A Special Report," available at: http://www.infinite-energy.com/images/pdfs/mitcfreport.pdf.

p. 65   *Mallove's report about the affair*: E. Mallove, *Ten Years That Shook Physics*, Infinite Energy, March-April, 1999.

p. 67   *the CR39 chip data*: *Naturwissenschaften* 94, no. 6 (2007): 511.

p. 67   *One of the few publications*: Quoted in Footlick, *Truth and Consequences*, p. 51.

5. LIFE

p. 70   *"What is life?"*: E. Schrödinger, *What is Life?* (Cambridge: Cambridge University Press, 1967), p. 3.

p. 71   *The physicist Paul Davies has perhaps done most*: P. Davies, *The Fifth Miracle* (London: Allan Lane, 1998), p. 7.

p. 71   *A living system must also be contained*: L. Margulis, D. Sagan, N. Eldredge, *What is Life?* (New York: Simon and Schuster, 1995), p. 113.

p. 71   *In June 2007 an editorial*: *Nature* 447 (2007): 1031.

p. 72   *In 1953 they sealed ammonia*: *Science* 130 (1959): 245.

p. 72   *Robert Shapiro likened the experiment's production*: R. Shapiro, "Where Do We Come From?" in *How Things Are*, ed. J. Brockman and K. Matson (London: Weidenfeld and Nicolson, 1995), p. 46.

p. 72   *Oro put water, hydrogen cyanide, and ammonia together*: *Nature* 191 (1961): 1193.

p. 72   *"life is either a reproducible"*: C. de Duve, *Vital Dust: Life as a Cosmic Imperative* (New York: Basic Books, 1996), p. 292.

p. 73   *Carl Sagan took the rapidity of life's emergence*: *Bioastronomy News* 7, no. 4 (1995).

p. 74   *"We knew the world"*: See http://www.atomicarchive.com/Movies/Movie8 .shtml.

p. 77   *Venter headed the team*: *Science* 270 (1995): 397.

p. 77   *"radically engineered organism"*: P. Aldhous, "Countdown to a Synthetic Life-form," *New Scientist*, July 11, 2007, p. 6.

p. 77   *"minimal cell project"*: *Anatomical Record* 268 (2002): 208.

p. 77   *At Harvard, Jack Szostak is also planning*: *Nature* 409 (2001): 387.

p. 78   *Anderson has always been a provocative voice*: *Science* 177 (1972): 393.

p. 79   *two more physicists took up Anderson's stance*: *Proceedings of the National Academy of Sciences* 97, issue 1 (2000): 28.

p. 80   *"organisms are not just tinkered-together contraptions"*: S. Kauffman, *At Home in the Universe* (New York: Oxford University Press, 1995).

p. 80   *"true source of physical law"*: R. Laughlin, *A Different Universe* (New York: Basic Books, 2005), p. 208.

p. 81   *Carl Sagan perhaps said it best*: "Visions of the Twenty-first Century," a speech given at St. John the Divine Cathedral in New York, 1995. Available at: http://www.atheistfoundation.org.au/carlsagan.htm.

p. 81   *"revel in our insignificance"*: G. Johnson, *Miss Leavitt's Stars* (New York: Atlas Books, 2005), p. 11.

p. 81   *The study took several years*: *Science* 274 (1996): 161.

6. VIKING

p. 84   *NASA researchers are drawing up work schedules*: See http://nssdc.gsfc.nasa .gov/planetary/mars/mars_colonize_terraform.html.

p. 89   *Levin counters this*: For a discussion of extremophiles, see M. Gross, *Life on the Edge* (New York: Perseus, 1998), p. 16.

p. 91   *Levin and Lafleur published*: "Instruments, Methods, and Missions for Astrobiology," *SPIE Proceedings* 4137 (2000): 48.

p. 92   *In 2006 the final nail was driven*: *Proceedings of the National Academy of Sciences* 103, no. 44 (2006): 16089.

p. 93   *"over 90 percent" certain*: L. Oliwenstein, "A Day in the Life on Mars," *University of Southern Califonia Health*, Winter 2002.

p. 93   *NASA's Chris McKay*: D. L. Chandler, "Searching for Life in a Handful of Dust," *New Scientist*, October 30, 2006, p. 48.

p. 94   *As you scroll through NASA's list*: See, for example, http://mars.jpl.nasa.gov/missions/.

p. 95   *Ward is unequivocal*: P. Ward, *Life as We Do Not Know It* (New York: Penguin Viking, 2005), p. 239.

p. 95   *Rees made the statement in a book*: M. Rees, "Cosmological Challenges: Are We Alone, and Where?" in *The Next Fifty Years*, ed. J. Brockman (London: Weidenfeld and Nicolson, 2002), p. 18.

p. 95   *Elsewhere he argued*: M. Rees, "Is the Search for Alien Life Futile Nonsense?" *New Scientist*, July 12, 2003, p. 25.

p. 96   *Piet Hut . . . has offered fifty-fifty odds*: See http://www.newscientist.com/article/dn10485-piet-hut-forecasts-the-future-.html.

p. 96   *Life's solutions are constrained by the laws of physics*: S. Conway Morris, *Life's Solution* (Cambridge: Cambridge University Press, 2003), p. 285.

7. THE WOW! SIGNAL

p. 97   *the likely characteristics of an alien communication*: Nature 184, no. 4690 (1959): 844.

p. 100   *We are living at a time of extraordinary progress*: Catalog at: http://exoplan-ets.org/planets.shtml.

p. 100   *scientists announced they had discovered three planets*: Nature 441 (2006): 305.

p. 101   *6EQUJ5 was the signature of a signal*: See http://www.bigear.org/6equj5.htm.

p. 102   *Occasionally something interesting*: See http://www.bigear.org/wow20th.htm.

p. 103   *He called it a day of infamy*: See http://www.bigear.org/JDK-Infamy.htm.

p. 108   *"You wouldn't believe cold fusion unless"*: See http://www.seti.org/about-us/faq.php.

8. A GIANT VIRUS

p. 113   *The bacterium was in fact not a bacterium. It was a giant virus*: Science 299 (2003): 2033.

p. 113   *Raoult subsequently admitted*: M. Peplow, "Giant Virus Qualifies as 'living' Organism," *Nature News Service*, October 14, 2004.

p. 115   *Woese published a paper*: Proceedings of the National Academy of Sciences 87, no. 12 (1977): 4576.

p. 117   *Mimivirus was proving to be a gold mine*: Science 306 (2004): 1344.

p. 118   *Bell came up with a rather surprising hypothesis*: *Journal of Molecular Evolution* 53 (2001): 251.

p. 119   *Mimivirus, he says, is the missing link*: G. Hamilton, "Half Virus, Half Beast," *New Scientist*, March 25, 2006, p. 37.

p. 119   *"the world's leading source of genetic innovation"*: L. Villarreal, "Are Viruses Alive?" *Scientific American*, September 2004, p. 96.

p. 120   *one-hundred-foot boat called* Sorcerer II: The project's Web page is http://www.sorcerer2expedition.org/version1/HTML/main.htm.

p. 120   *around 10 percent of them*: *Emerging Infectious Diseases* 11 (2005): 449.

p. 120   *a study in France*: *Microbial Pathogenesis* 42 (2007): 56.

p. 120   *a technician in the Marseille lab*: *Annals of Internal Medicine* 144 (2006): 702.

p. 121   *It is called a* reovirus: *Science* 282 (1998): 1332.

9. DEATH

p. 122   *a young researcher from the University of Georgia*: W. Gibbons, "How Long Do Blanding's Turtles Live?" *Ecoviews*, http://www.uga.edu/srelherp/ecoview/Eco25.htm.

p. 123   *"a sharp challenge"*: B. Yeoman, "Can Turtles Live Forever?" *Discover* magazine, January 6, 2002, p. 61.

p. 124   *"sheer, wanton, head-in-bag perversity"*: *Behavioral and Brain Sciences* 17, no. 4 (1994): 616.

p. 124   *With great insight, he proposed a mechanism*: P. Medawar, *An Unsolved Problem in Biology* (London: H. K. Lewis, 1952), p. 1.

p. 124   *In 1957 George Williams expanded*: *Evolution* 11 (1957): 398.

p. 124   *Then, in 1977, Tom Kirkwood*: *Nature* 270 (1977): 301.

p. 125   *"highly controversial"*: BBC Reith Lectures, 2001, transcript at: http://www.bbc.co.uk/radio4/reith2001/lecture3.shtml.

p. 125   *Thomas Johnson and David Friedman joined*: *Genetics* 118 (1988): 75.

p. 125   *some of their colleagues accused them*: G. Hamilton, "Clock of Ages," *New Scientist*, April 19, 2003, p. 26.

p. 125   Caenorhabditis elegans *worms . . . up to six weeks*: *Nature* 366 (1993): 461.

p. 126   *"organism envy"*: *Science* 308 (2005): 1875.

p. 126   *The group was headed . . . fifty-one scientists*: *Journals of Gerontology Series A: Biological Sciences and Medical Sciences* 57 (2002): B292–B297.

p. 126   *"No intervention will slow"*: *Journals of Gerontology Series A: Biological Sciences and Medical Sciences* 59 (2004): B573–B578.

p. 128   *"Our predictions were met with disbelief"*: *Citation Classics*, no. 26 (1978):

144, available at: http://www.garfield.library.upenn.edu/classics1978/A1978
FC39200002.pdf.

p. 128 *they had put a gene that activates telomerase*: Science 279 (1998): 349.

p. 129 *a tantalizing secret*: Nature 448 (2007): 767.

p. 129 *their fertility declined*: Evolution 38 (1980): 1004.

p. 130 *increased life span and increased fertility*: Evolution 45 (1991): 82.

p. 130 *female mice shut down*: Science 190 (1975): 165.

p. 130 *As her group point out in a 2003 paper*: Science 302 (2003): 611.

p. 131 *His conclusion was that there was no conclusion*: Evolutionary Ecology Research
6 (2004): 1.

p. 132 *"Biological Aging Is No Longer an Unsolved Problem"*: Annals of the New
York Academy of Sciences 1100 (2007): 1.

p. 133 *common ancestor of today's species*: W. A. Clark, *Means to an End: The Biological Basis of Aging and Death* (New York: Oxford University Press,
1991), p. 41.

10. SEX

p. 136 *an outstanding exposition of the theory*: R. Dawkins, *Climbing Mount Improbable* (New York, Norton, 1996), p. 75.

p. 136 *he again admits defeat*: R. Dawkins, *The Ancestor's Tale* (London: Weidenfeld
and Nicolson, 2004), p. 357.

p. 136 *"evolutionary scandal"*: J. Maynard Smith, Nature 324 (1986): 300.

p. 137 *"a kind of crisis at hand"*: G. Williams, *Sex and Evolution* (Princeton: Princeton University Press, 1975), p. v.

p. 137 *Ernst Mayr added his contribution*: E. Mayr, *What Evolution Is* (London: Weidenfeld and Nicolson, 2002), p. 102.

p. 137 *Bringing things right up to date*: Nature Reviews (Genetics) 8 (2007): 139.

p. 137 *"twofold cost"*: Maynard Smith, *The Evolution of Sex* (Cambridge: Cambridge
University Press, 1978), p. 3.

p. 138 *turned this argument upside down*: Science 288, no. 5469 (2000): 1211.

p. 139 *Autumn's asexual geckos . . . farther and faster*: Physiological and Biochemical
Zoology 78 (2005): 3.

p. 139 *A series of experiments on water fleas*: This and other results discussed in this
paragraph are summarized in Nature Reviews (Genetics) 8 (2007): 139.

p. 140 *Graham Bell and Austin Burt showed*: Nature 330 (1987): 118.

pp. 140 *"a good candidate for the title"*: R. Dawkins, Independent, March 10, 2000.
–141

p. 141   *Water fleas have shown no advantage*: *Journal of Evolutionary Biology* 16 (2003): 976.

p. 141   *for rotifers, their advantage lies*: *Science* 18 (2007): 268.

p. 141   *In 2004 Sarah Otto and Scott Nuismer struck another blow against the Red Queen*: *Science* 304 (2004): 1018.

p. 143   *"hidden in darkness"*: *Journal of the Proceedings of the Linnean Society of London (Botany)* 6: 95.

p. 143   *More than a century later*: Maynard Smith, *The Evolution of Sex*.

p. 143   *something Kuhn called "a scandal"*: T. Kuhn, *The Structure of Scientific Revolutions* (Chicago: University of Chicago Press, 1962), p. 67.

p. 144   *In some ways the issue parallels*: D. Gale and L. Shapley, *American Mathematical Monthly* 69 (1962): 9.

p. 145   *wholesale replacement of Darwin's theory*: *Science* 311 (2006): 965.

p. 146   *it still stands as a point of contention*: *Evolution* 59 (2005): 87.

p. 147   *As the biologist Steven Rose pointed out*: S. Rose, "Chat-Up Lines," *Guardian*, August 21, 2004.

p. 147   *In the summer of 1994*: *Proceedings of the Royal Society of London* 264 (1997): 1283.

p. 148   *more than 450 species*: B. Bagemihl, *Biological Exuberance: Animal Homosexuality and Natural Diversity* (London: Profile Books, 1999), p. 12.

p. 148   *she took the total number of vertebrate species observed*: J. Roughgarden, *Evolution's Rainbow* (Berkeley: University of California Press, 2004), p. 224.

p. 149   *Steven Rose wrote*: Rose, "Chat-Up Lines."

p. 150   *Jerome Wodinsky removed*: *Science* 198 (1997): 880.

11. FREE WILL

p. 152   *There are plenty of other examples*: O. Sacks, *The Man Who Mistook His Wife for a Hat and Other Clinical Tales* (New York: Summit Books, 1985), p. 8.

p. 152   *"Man defends himself from being regarded"*: Quoted in *Journal of Consciousness Studies* 2, no. 2 (1995): 167.

p. 153   *In 1788 the philosopher Immanuel Kant*: I. Kant, *Critique of Practical Reason*, ed. and translated by L. Beck (Cambridge: Cambridge University Press, 1997), p. 2.

p. 153   *Libet found that the brain's preparatory work*: *Brain* 106 (1983): 623.

p. 154   *That was certainly Libet's view*: B. Libet, "Do We Have Free Will?" in *The Volitional Brain*, ed. B. Libet, A. Freeman, and V. Sutherland (Exeter: Imprint Academic, 1999), p. 47.

p. 156   *Fried grasped this opportunity*: *Journal of Neuroscience* 11 (1991): 3656.

p. 158   *The students gained a course credit*: "Apparent Mental Causation," *American Psychologist*, July 1999, p. 480.

p. 158   *In these studies, the students*: *Journal of Personality and Social Psychology* 85, no. 1 (2003): 5.

p. 159   *"influence of suggestion"*: *Royal Institution of Great Britain (Proceedings)*, March 12, 1852, p. 147.

p. 159   *William James . . . took Carpenter's baton*: W. James, *The Principles of Psychology* (New York: H. Holt, 1890), p. 526.

p. 160   *This, perhaps, is what is most disturbing*: A. Burgess, *A Clockwork Orange* (London: Heinemann, 1962), p. 76.

p. 162   *Richard Nisbett and Timothy Wilson showed*: *Psychological Review* 84 (1977): 231.

p. 163   *"Free will is a fictional construction"*: "In search of humanity," *Times* (London), December 29, 1997.

12. THE PLACEBO EFFECT

p. 164   *"It has brought me great comfort"*: Press release on Sternbach's death issued by Roche Pharmaceuticals, September 30, 2005.

p. 164   *Diazepam is now . . . a "core medicine"*: See http://www.who.int/medicines/publications/EML15.pdf.

p. 165   *diazepam had no effect on anxiety*: *Prevention and Treatment* 6, no. 1 (2003): v.

p. 166   *"urgent priority"*: Quoted in L. Conboy et al., *Contemporary Clinical Trials* 27 (2006): 123.

p. 167   *"I'm going to prescribe you some magnesium"*: L. Spinney, "Purveyors of Mystery," *New Scientist*, December 16, 2006, p. 42.

p. 167   *"some unintelligent or inadequate patients"*: *Lancet* 2 (1954): 321.1.

p. 167   *According to Ann Helm*: A. Helm, "Truth-Telling, Placebos and Deception," *Aviation, Space, and Environmental Medicine*, January 1985, p. 69.

p. 167   *Danish clinicians . . . ten or more times per year*: *Evaluation and the Health Professions* 26 (2003): 153.

p. 167   *Israeli doctors . . . prescribed placebos*: *British Medical Journal* 329 (2004): 944.

p. 168   *"Generally a larger dose"*: *Journal of the American Pharmaceutical Association* 41, no. 4 (2001): 523.

p. 170   *Asbjorn Hróbjartsson and Peter Gøtzsche had begun*: *New England Journal of Medicine* 344 (2001): 1594.

p. 170   *much-quoted, never-questioned statistic*: Journal of the American Medical Association 159 (1955): 1602.

p. 171   *In 2003 Hróbjartsson and Gøtzsche*: Journal of Internal Medicine 256 (2003): 91.

p. 171   *researchers from the University of Michigan*: Journal of Neuroscience 25 (2005): 7754.

p. 172   *An editorial accompanying*: New England Journal of Medicine 344 (2001): 1630.

p. 174   *reduced activity in the neurons*: Nature Neuroscience 7 (2004): 587.

p. 175   *Telling patients . . . as effective as injecting 6–8 mg of morphine*: Nature 312 (1984): 755.

p. 175   *cocaine abusers . . . getting something*: Journal of Neuroscience 23 (2003): 11461.

p. 175   *Benedetti and Colloca have already started*: Nature Reviews (Neuroscience) 6 (2005): 545.

p. 176   *his team published a paper*: Journal of Neuroscience 26 (2006): 12014.

p. 178   *One group, led by researchers*: Contemporary Clinical Trials 27 (2006): 123.

p. 178   *An openly administered dose*: Pain 90 (2001): 205.

13. HOMEOPATHY

p. 181   *According to the World Health Organization*: Bulletin of the World Health Organization 77 (1999): 160.

p. 183   *Benveniste convinced the journal* Nature: Nature 333 (1988): 816.

p. 183   Nature *published a critique*: Nature 334 (1988): 291.

p. 183   *"incredibly surprised"*: L. Milgrom, "Thanks for the Memory," Guardian, March 15, 2001.

p. 184   *The trial . . . took place in four different laboratories*: Inflammation Research 50 (2001): 47.

p. 185   *a team of scientists failed to replicate*: See http://www.bbc.co.uk/science/horizon/2002/homeopathy. shtml.

p. 185   *she later distanced herself*: See http://www.homeopathic.com/articles/view,55.

p. 185   *A study by Adrian Guggisberg*: Complementary Therapies in Medicine 13, no. 2 (2005): 91.

p. 186   *Dylan Evans attributes . . . the placebo effect*: D. Evans, Placebo (London: Harper Collins, 2003), p. 149.

p. 186   *a 1997 meta-analysis published in the* Lancet: Lancet 350 (1997): 834.

p. 186   *Robert L. Park uses the same argument*: R. Park, Voodoo Science (New York: Oxford University Press, 2000), p. 57.

p. 187   *At least sixty-four*: See http://www.lsbu.ac.uk/water/.

p. 188   *His review article reads like a political address*: Nature Reviews (Molecular Cell Biology) 7, no. 11 (2006): 861.

p. 189   *a group of German researchers*: Angewandte Chemie (international edition) 40 (2001): 1808.

p. 189   *it becomes broken up into distinct beads*: Journal of Chemical Physics 120 (2004): 5867.

p. 189   *Anders Nilsson published a paper*: Science 304 (2004): 995.

p. 191   *Roy advocates using silver as an antibiotic*: Materials Research Innovations 11, no. 1 (2007): 3.

p. 191   *repeatedly separated fools and their money*: The FDA regulation on the subject is at: http://a257.g.akamaitech.net/7/257/2422/10apr20061500/edocket.access.gpo.gov/cfr_2006/aprqtr/pdf/21cfr310.548.pdf.

p. 192   *Its editorial article*: Lancet 366 (2005): 690.

p. 192   *an article published in the same issue*: Ibid., p. 726.

p. 192   *"flawed" study*: Ibid., p. 2081.

p. 192   *Klaus Linde and Wayne Jonas had published*: Lancet 350 (1997): 834.

p. 193   *But then Jonas*: Journal of Alternative and Complementary Medicine 11, no. 5 (2005): 813.

p. 194   *A six-month study*: Rheumatology 40 (2001): 1052.

p. 201   *This dangerous ideal . . . "posthumous ostracism"*: S. Land, "The Two Faces of Homoeopathy," British Homoeopathic Journal, January 1985, p. 49.

EPILOGUE

p. 204   *J. Donald Fernie made a wry observation*: Publication of the Astronomical Society of the Pacific 81 (1969): 707.

p. 205   *Horgan argued that science is, essentially, finished*: J. Horgan, The End of Science (Reading, Mass: Addison Wesley, 1996), p. 1.

p. 206   *Simon White has, for example, suggested*: Reports on Progress in Physics 70 (2007): 883.

p. 208   *The widespread acceptance . . . is unwarranted*: R. Laughlin, A Different Universe (New York: Basic Books, 2005), p. 211.

p. 209   *very young or very new*: T. Kuhn, The Structure of Scientific Revolutions (Chicago: University of Chicago Press, 1962), p. 151.

p. 210   *"Who are we?"*: E. Schrödinger, Science and Humanism (Cambridge: Cambridge University Press, 1951).

INDEX

End of Science, The (Horgan),
205–6
endorphins, 169, 171
energy, 5, 25–26, 134
    connection of mass and, 8–9, 10,
        19, 26
    conservation of, 2, 28
    quantum packets of, 6, 207
    See also specific forms of energy
Energy Department.
    See Department of Energy, U.S.
Energy Research Advisory Board
        (ERAB), 60, 61, 64, 65
Ennis, Madeleine, 183–85
entropy, 70
epilepsy, 156, 157
epistasis, 140
epitaxy, 190
erythrocytes, 97
ether, 34
ethology, 79
eukaryotes, 115, 116, 118, 119, 121,
        133–34, 140, 143–44, 149
Europa, 100
European Southern Observatory,
        54
Evaluation and the Health Professions,
        167
Evans, Dylan, 186
evolution, 71, 128, 209
    natural selection and, 6, 114,
        123–24, 131–32, 137, 138, 145
Evolutionary Ecology Research, 131
Evolution's Rainbow (Roughgarden),
        148–49
extra universal force, 43
Extravagant Universe, The
        (Kirshner), 23

F

Fermi, Enrico, 107
Fernie, J. Donald, 204
Feynman, Richard, 56
Fifth Miracle, The (Davies), 71
final theory, 8, 205–6
Finch, Caleb, 123
Fleischmann, Martin, 56, 57–62,
        63–65, 67–68, 189, 203
Forbes, John, 182
Forsgren, Elisabet, 147, 148
Fox, George, 115
Franklin, Benjamin, 166, 177
Franklin, Rosalind, 186–87
free radicals, 130
free will, 5, 150–63
    conscious, 159, 162
    experiments in, 151, 153–60
    illusion of, 152–54, 155–56,
        158–60, 162–63, 209, 210
French atomic energy commission,
        52–53
Fried, Itzhak, 156, 157
Friedman, David, 125
fruit flies, 126–27, 129–30, 131, 147
Fuller, Buckminster, 188

G

Gagarin, Yuri, 39
galaxies, 4, 17, 47, 81
    centers vs. outsides of, 15
    clusters of, 13–14, 29, 31–32
    collision of, 31–32
    distances between, 8, 29
    distribution of mass in, 14, 16, 22
    gravitational influence on, 16, 29

molecules, 70, 75, 76, 93, 113, 187–89
MOND (Modified Newtonian
    Dynamics), 29–32, 43
Moon, 36, 88
    Apollo landings on, 9, 37
    craters on, 11, 73
moons, 36, 81, 100, 204, 206
Moorhead, Paul, 127–28
"More Is Different" (Anderson), 78,
    79
Morley, Edward, 205
morphine, 169–70, 175
Morrison, Philip, 97–98, 101, 107,
    108
Motherby, George, 166
Mount Wilson Observatory, 14
mu, 54–55
Muller's ratchet, 139
Mycoplasma genitalium, 77

N

naloxone, 169–70
NASA (National Aeronautics and
    Space Administration), 33,
    37–38, 42, 81, 85, 103
    Jet Propulsion Laboratory (JPL) of,
        38, 41
    Mars missions of, 83, 84, 85–95, 99
    Microwave Observing Program
        (MOP) of, 104
    proof of dark matter announced
        by, 31–32
National Academy of Sciences, 11
National Institutes of Health, 166,
    175
National Science Foundation, 33, 80

National Security Agency, U.S., 98
Nature, 25, 63, 71, 97, 98, 125, 129,
    137, 183, 187–88
Naval Air Warfare Center, 61
Naval Observatory, U.S., 15
Navy, U.S., 64
    Office of Naval Research of, 61, 62,
        66
NCLDV (nucleocytoplasmic large
    DNA virus), 113–14, 117
nebulae, 9–12
Nedelcu, Aurora, 139
Neptune, 13, 37, 43, 44, 84
Neumann, John von, 2, 145
neuroscience, 154–57, 161–63, 174
neutralinos, 17–18, 32
neutrons, 59–60, 79
Newcomb, Simon, 205
New Medical Dictionary (Hooper),
    166
New Medical Dictionary (Motherby),
    166
New Pathways in Science
    (Eddington), 53–54
Newton, Isaac, 4, 10, 28–33, 36–38,
    41, 50, 132
    See also gravity; MOND
New York Times, 63, 105, 106
Nieto, Michael Martin, 38, 40, 42, 43
Nilsson, Anders, 189
Nisbett, Richard, 162
nitrogen monoxide, 129
Nobel Prize for Physics, 6, 50, 55, 56,
    58, 206
nocebo effect, 175–76
Nowak, Robert, 62
nuclear fission, 66
nuclear fusion, 49, 57, 58–59, 66

Type 1a, 21
Susskind, Leonard, 27
Systema natura (Linnaeus), 114–15
Szostak, Jack, 77

T

Taurus constellation, 37
telescopes, 21, 23–24, 48, 51, 54, 84, 102
    See also specific telescopes
telomerase, 128
telomeres, 119, 128–29
10 Years That Shook Physics (Mallove), 65
theory of everything, 3
Tiller, William, 190, 191
time, 6, 10, 39
Titan, 100
tobacco mosaic virus, 11
Tomonaga, Shin'ichiro, 56
Torgerson, Justin, 53
transcranial magnetic stimulation, 151
tritium, 59
Turner, Michael, 22–23
Turyshev, Slava, 38–41, 42, 43, 44

U

Ultraphoton, 39
uncertainty principle, 27–28, 177
unification theory, 54
universe, 7–35
    accelerated expansion of, 3, 8, 12–13, 19–20, 21, 22, 23, 24–25, 26, 42, 43, 49, 207, 208

age of, 23
constituent particles and forces of, 8–9, 10
evolution of, 35, 56
four fundamental forces in, 50, 53–54
gravitational pull on, 13–14, 191, 22–23
mystery of 96 percent of, 4, 8, 13
Omega value of, 20, 22–24
space and time as fabric of, 10, 28
subuniverses in, 27
varied terrain and properties of, 27, 28
    See also big bang
Universe in a Nutshell, The (Hawking), 12
uranium, 52–53
Uranus, 84
Urey, Harold C., 72

V

vacuum energy, 26
Valium, 164
varicella zoster, 113
Venter, Craig, 76–78, 81, 120
Venus, 81, 84, 107
vertebrates, 122–23
Viking missions, 84, 85–94
    experiments of, 86–87, 89, 92, 93
Villarreal, Luis, 119–20
Virginia, University of, 158
viruses, 110–21, 129
    antibodies to, 120
    death and, 119, 121, 124
    discovery of, 111, 113